Encyclopedia of Alternative and Renewable Energy: Advanced Thin Film Solar Cell Techniques

Volume 25

Encyclopedia of Alternative and Renewable Energy: Advanced Thin Film Solar Cell Techniques Volume 25

Edited by **Terence Maran and David McCartney**

New York

Published by Callisto Reference,
106 Park Avenue, Suite 200,
New York, NY 10016, USA
www.callistoreference.com

Encyclopedia of Alternative and Renewable Energy:
Advanced Thin Film Solar Cell Techniques
Volume 25
Edited by Terence Maran and David McCartney

International Standard Book Number: 978-1-63239-199-5 (Hardback)

Contents

Permissions

List of Contributors

Preface

This book aims to highlight the current researches and provides a platform to further the scope of innovations in this area. This book is a product of the combined efforts of many researchers and scientists, after going through thorough studies and analysis from different parts of the world. The objective of this book is to provide the readers with the latest information of the field.

This is an advanced book on thin film solar cell techniques. This technique uses direct-gap semiconductors such as CIGS and CdTe pose minimum manufacturing costs and is now increasing in popularity amongst industries. The field of photovoltaics has seen a large-scale manufacturing of the second genesis of thin film solar modules and has succeeded in constructing powerful solar plants in many countries across the globe. This has led to an increase in the manufacturability of thin film solar modules as compared to wafer or ribbon Si modules. Thin films like CIGS and CdTe will soon take over wafer-based silicon solar cells as the superior photovoltaic technology. This book elucidates the scientific and technological difficulties of increasing the photoelectric efficiency of thin film solar cells. It covers various aspects of thin film solar cells processing, modeling and sensitive issues, analysis of monograin layer solar cell etc. The book will be beneficial for readers interested in this subject.

I would like to express my sincere thanks to the authors for their dedicated efforts in the completion of this book. I acknowledge the efforts of the publisher for providing constant support. Lastly, I would like to thank my family for their support in all academic endeavors.

Editor

Chemical Bath Deposited CdS for CdTe and Cu(In,Ga)Se$_2$ Thin Film Solar Cells Processing

M. Estela Calixto[1], M. L. Albor-Aguilera[2], M. Tufiño-Velázquez[2],
G. Contreras-Puente[2] and A. Morales-Acevedo[3]

[1]Instituto de Física, Benemérita Universidad Autónoma de Puebla, Puebla,
[2]Escuela Superior de Física y Matemáticas, Instituto Politécnico Nacional, México,
[3]CINVESTAV-IPN, Departamento de Ingeniería Eléctrica, México,
México

1. Introduction

Extensive research has been done during the last two decades on cadmium sulfide (CdS) thin films, mainly due to their application to large area electronic devices such as thin film field-effect transistors (Schon et al., 2001) and solar cells (Romeo et al., 2004). For the latter case, chemical bath deposited (CBD) CdS thin films have been used extensively in the processing of CdTe and Cu(In,Ga)Se$_2$ solar cells, because it is a very simple and inexpensive technique to scale up to deposit CdS thin films for mass production processes and because among other n-type semiconductor materials, it has been found that CdS is the most promising heterojunction partner for these well-known polycrystalline photovoltaic materials. Semiconducting n-type CdS thin films have been widely used as a window layer in solar cells; the quality of the CdS-partner plays an important role into the PV device performance. Usually the deposition of the CdS thin films by CBD is carried out using an alkaline aqueous solution (high pH) composed mainly of some sort of Cd compounds (chloride, nitrate, sulfate salts, etc), thiourea as the sulfide source and ammonia as the complexing agent, which helps to prevent the undesirable homogeneous precipitation by forming complexes with Cd ions, slowing down thus the surface reaction on the substrate. CdS films have to fulfill some important criteria to be used for solar cell applications; they have to be adherent to the substrate and free of pinholes or other physical imperfections. Moreover, due to the requirements imposed to the thickness of the CdS films for the solar cells, it seems to be a function of the relative physical perfection of the film. The better structured CdS films and the fewer flaws present, the thinner the film can be, requirement very important for the processing of Cu(In,Ga)Se$_2$ based thin film solar cells, thickness ~ 30 - 50 nm. In such case, the growth of the thin CdS film is known to occur via ion by ion reaction, resulting thus into the growth of dense and homogeneous films with mixed cubic/hexagonal lattice structure (Shafarman and Stolt, 2003).

The reason to choose the CBD method to prepare the CdS layers was due to the fact that CBD forms a very compact film that covers the TCO layer, in the case of the CdTe devices and the Cu(In,Ga)Se$_2$ layer without pinholes. Moreover, the CdS layer in a hetero-junction solar cell must also be highly transparent and form a chemical stable interface with the

Cu(In,Ga)Se$_2$ and CdTe absorbing layers. The micro-crystalline quality of the film may also be related to the formation of the CdZnS ternary layer in the case of the Cu(In,Ga)Se$_2$ and CdS$_{1-x}$Te$_x$ ternary layer for the case of CdTe, at the interface helping to reduce the effects associated to the carrier traps in it. Hence, the deposition conditions and characteristics of the CdS layer may affect strongly the efficiency of the solar cells. We have worked with this assumption in mind for making several experiments that will be described in the following paragraphs. As it will be shown, we have been able to prepare optimum CdS layers by CBD in order to be used in solar cells, and have found that the best performance of CdS/CdTe solar cells is related to the CdS layer with better micro-crystalline quality as revealed by photoluminescence measurements performed to the CdS films.

2. CdS thin films by chemical bath deposition technique (CBD)

Chemical bath deposition technique (CBD) has been widely used to deposit films of many different semiconductors. It has proven over the years to be the simplest method available for this purpose, the typical components of a CBD system are a container for the solution bath, the solution itself made up of common chemical reactive salts, the substrate where the deposition of the film is going to take place, a device to control the stirring process and temperature, sometimes a water bath is included to ensure an homogeneous temperature, an schematic diagram of the CBD system is shown in figure 1. The concentrations of the components of the solution bath for CdS can be varied over a working range and each group use its own specific recipe, so there are as many recipes to deposit CdS as research groups working in the subject. The chemical reactive salts are generally of low cost and in general it is necessary to use small quantities. The most important deposition parameters in this technique are the molar concentration, the pH, the deposition temperature, the deposition time, the stirring rate, the complexing agents added to the bath to slowing down the chemical reactions, etc. However, once they have been established these are easy to control. The CdS thin film deposition can be performed over several substrates at a time, and the reproducibility is guaranteed if the deposition parameters are kept the same every time a deposition is done. Substrates can have any area and any configuration, besides they can be of any kind, electrical conductivity is not required.

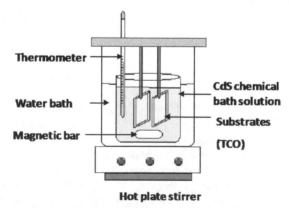

Hot plate stirrer

Fig. 1. Schematic diagram of a CdS chemical bath deposition system

Previously we have reported the preparation of monolayers and bi-layers of CdS deposited by chemical bath deposition technique using a solution bath based on CdCl$_2$ (0.1 M), NH$_4$Cl (0.2 M), NH$_3$ (2 M) and thiourea (0.3 M), maintaining fixed deposition time and temperature conditions and varying the order of application of the CdCl$_2$ treatment (Contreras-Puente et al., 2006). Initially, the solution is preheated during 5 min prior to add the thiourea, after that the deposition was carried out during 10 min at 75 C, then the second layer (the bi-layer) was deposited at a lower deposition temperature, thus allowing us to control the growth rate of the CdS layer. This was aimed to obtain films with sub-micron and nanometric particle size that could help to solve problems such as partial grain coverage, inter-granular caverns and pinholes. In this way, CdS thin films have been deposited onto SnO$_2$: F substrates of 4 cm^2 and 40 cm^2, respectively.

Figure 2 shows the typical X-ray diffraction pattern obtained with a glancing incidence X-ray diffractometer, for CdS samples prepared in small and large area, respectively. CdS films grow with preferential orientation in the (002), (112) y (004) directions, which correspond to the CdS hexagonal structure (JCPDS 41-049). Small traces of SnO$_2$:F are observed (*) in the X-ray patterns. Figure 3 shows the morphology for both mono and bi-layers of CdS films, respectively. It can be observed that bi-layer films present lower pinhole density and caverns. This is a critical parameter because it gives us the possibility to improve the efficiency of solar cell devices. Several sets of CdTe devices were made and their photovoltaic parameters analyzed, giving conversion efficiencies of ~ 6.5 % for both small and large area devices.

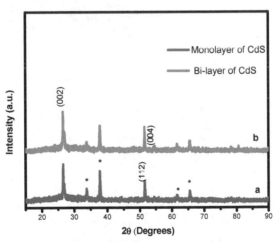

Fig. 2. X-ray diffraction patterns of mono and bi-layers of CdS

Also, we have found that the position of the substrate inside the reactor is an important factor because the kinetics of the growth changes. Figure 4 shows how the transmission response changes with substrate position inside the reactor. The deposition time for all samples was 10 min. According to figure 4a when the substrates are placed horizontally at the bottom of the reactor the CdS film grows a thickness of 150 nm, but the transmission response is poor, when the substrates are placed vertically and suspended with a pair of tweezers inside the reactor the CdS film grows a thickness of 110 nm and the transmission response is ~ 83% (see figure 4b), however in this configuration handling the substrate is complicated. Because of this, to design a better substrate holder/support was imperative. So, a new support was designed and built to facilitate the access and handling of the samples inside the reactor. Figures 4c and 4d shown the transmission response for mono and bi-layers of CdS deposited using the new substrate support, placed in a vertical configuration inside the reactor, for both cases the values were between 85 – 95 %, being the monolayers the ones that exhibit the best response; however its morphology shows a larger surface defect density. The thickness of these samples is in the order of 100 – 120 nm.

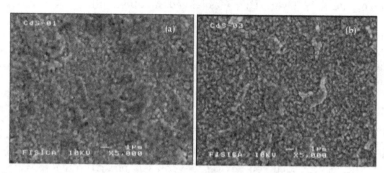

Fig. 3. SEM images of a monolayer and a bi-layer of CdS

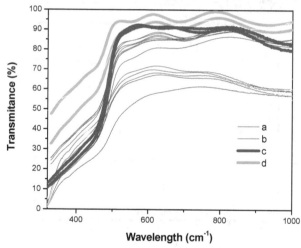

Fig. 4. Transmission response of CdS films as a function of the position inside the reactor

2.1 CdS by CBD with a modified configuration

Figure 5 shows the implementation of the new substrate support for the CBD system, from this figure it can be seen that the CBD system is the same as the one shown in figure 1 but with the addition of the substrate holder. It basically holds the substrates vertically and steady, while keeping it free to rotate along with the substrates, in such case the magnetic stirrer is no longer needed. This substrate support can be set to rotate at different speed rates, allowing the growth and kinetics of the reaction of CdS to change and in the best case to improve, improving thus the physical properties of CdS films. The design includes a direct current motor that has the option to vary the speed rate from 0 to 50 rpm. The motor can move the substrate support made of a Teflon structure that holds up to 4 large area substrates (45 cm² each). The principal advantage of using this modified structure is the ability to handle 4 substrates at a time, placing them, inside the reactor containing the solution bath and at the same time starting the rotation, by doing this all the CdS films are expected to have a uniform growth and thickness ~ 120 nm. When the substrate holder is set to rotate inside the reactor, the kinetics of the CdS films growth was clearly affected as shown in figure 6, it can be seen that when the rotating speed goes up, the transmission

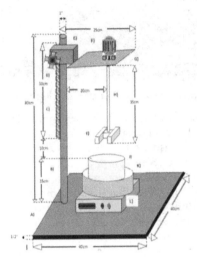

Fig. 5. Schematic diagram of the new substrate holder for the CBD system

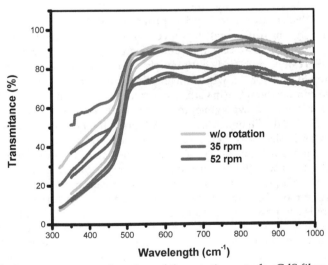

Fig. 6. Transmission response as a function of the rotation rate for CdS films prepared with the new substrate holder.

response decreases to ~ 65% compared to the samples prepared without rotation. The deposition time was set to 10 min in all cases, giving thus the growth of CdS films with 120 – 130 nm.

Fig. 7. SEM images of (a) mono and (b) bi-layer of CdS deposited at 35 rpm

Figure 7 shows the SEM images of CdS films prepared using the new substrate holder, according to these images, the morphology of the mono and bi-layers of CdS changes as a function of the rotating speed. Also we can clearly see an increase in the particle size for each case, for the monolayer of CdS the particle size ball- like shape of ~ 0.5 – 1 µm, but more uniform and compact compared to the particle size that the bi-layers of CdS exhibit with rotation speed set to 35 rpm, flakes-like shape with size of ~1 – 4 µm. No devices have been made so far using CdS films grown with this improved CBD system, studies are being performed and research on the subject is ongoing in order to optimize the deposition conditions, for this case.

3. Cu(In,Ga)Se$_2$ based thin films by co-evaporation technique (PVD)

Semiconducting CuInSe$_2$ is one of the most promising materials for solar cells applications because of its favorable electronic and optical properties including its direct band gap with high absorption coefficient (~10^5 cm^{-1}) thus layers of only ~2 µm thickness are required to absorb most of the usable solar radiation and inherent p-type conductivity. Besides, the band gap of CuInSe$_2$ can be modified continuously over a wide range from 1.02 to 2.5 eV by substituting Ga for In or S for Se, which means that this material can be prepared with a different chemical composition. Cu(In,Ga)Se$_2$ is a very forgiving material so high efficiency devices can be made with a wide tolerance to variations in Cu(In,Ga)Se$_2$ composition (Rocheleau et al., 1987 and Mitchell K. et al., 1990), grain boundaries are inherently passive so even films with grain sizes less than 1 µm can be used, and device behavior is insensitive to defects at the junction caused by a lattice mismatch or impurities between the Cu(In,Ga)Se$_2$ and CdS. The latter enables high-efficiency devices to be processed despite exposure of the Cu(In,Ga)Se$_2$ to air prior to junction formation. For Cu(In,Ga)Se$_2$ thin film solar cells processing the substrate structure is preferred over the superstrate structure. The substrate structure is composed of a soda lime glass substrate, coated with a Mo layer used as the back contact where the Cu(In,Ga)Se$_2$ film is deposited. The soda lime glass, which is used in conventional windows, is the most common substrate material used to deposit Cu(In,Ga)Se$_2$ since it is available in large quantities at low cost. Besides, it has a thermal expansion coefficient of 9×10^{-6} K^{-1} (Boyd et al., 1980) which provides a good match to the Cu(In,Ga)Se$_2$ films. The most important effect of the soda lime glass substrate on

Cu(In,Ga)Se₂ film growth is that it is a natural source of sodium for the growing material. So that, the sodium diffuses through the sputtered Mo back contact, which means that is very important to control the properties of the Mo layer. The presence of sodium promotes the growth of larger grains of the Cu(In,Ga)Se₂ and with a higher degree of preferred orientation in the (112) direction. After Cu(In,Ga)Se₂ deposition, the junction is formed by depositing a CdS layer. Then a high-resistance (HR) ZnO and a doped high-conductivity ZnO:Al layers are subsequently deposited. The ZnO layer reacts with the CdS forming the Cd$_x$Zn$_{1-x}$S ternary compound, which is known to have a wider band gap than CdS alone, increasing thus the cell current by increasing the short wavelength (blue) response and at the same time setting the conditions to make a better electric contact. Finally, the deposition of a current-collecting Ni/Al grid completes the device. The highest conversion efficiency for Cu(In,Ga)Se₂ thin film solar cells of ~ 20 % has been achieved by (Repins et al., 2008) using a three stages co-evaporation process. The processing of photovoltaic (PV) quality films is generally carried out via high vacuum techniques, like thermal co-evaporation. This was mainly the reason, we have carried out the implementation and characterization of a thermal co-evaporation system with individual Knudsen cells MBE type, to deposit the Cu(In,Ga)Se₂ thin films (see figure 8). The deposition conditions for each metal source were established previously by doing a deposition profile of temperature data vs. growth rate. The thermal co-evaporation of Cu(In,Ga)Se₂ thin films was carried out using Cu shots 99.999%, Ga ingots 99.9999%, Se shots 99.999% from Alfa Aeser and In wire 99.999% from Kurt J. Lesker, used as received. The depositions were performed on soda lime glass substrates with sputtered Mo with ~ 0.7 μm of thickness. The substrate temperature was > 500 °C, temperature of source materials was set to ensure a growth rate of 1.4, 2.2 and 0.9 Å/s for Cu, In and Ga, respectively for the metals, while keeping a selenium overpressure into the vacuum chamber during film growth.

Fig. 8. Thermal co-evaporation system with Knudsen effusion cells to deposit Cu(In,Ga)Se₂ thin films

$Cu(In,Ga)Se_2$ thin films were grown with different Ga and Cu ratios $(Ga/(In+Ga) = 0.28$, 0.34 and 0.35 respectively and $Cu/(In+Ga) = 0.85$, 0.83 and 0.94). The deposition time was set to 30 min for all cases. All the $Cu(In,Ga)Se_2$ samples were grown to have 2 - 3 μm thickness and aiming to obtain a relative low content of gallium ~ 0.30 % $(CuIn_{0.7}Ga_{0.3}Se_2)$, while keeping the copper ratio to III < 1 (where III = In+Ga), very important criteria to use them directly for solar cell applications, as shown in table 1. For solar cell devices, samples JS17 and JS18 were used, with a chemical composition similar to that of sample JS13.

Chemical composition (at %) by EDS						
Sample	Cu	In	Ga	Se	Ga/III	Cu/III
Reference	22.09	18.84	7.27	51.80	0.28	0.85
CIGS_5	21.27	16.73	8.88	53.69	0.35	0.83
CIGS_8	23.04	16.20	8.24	53.47	0.34	0.94
JS13	24.46	16.87	9.74	48.93	0.37	0.92

Table 1. Results of the chemical composition analysis of the co-evaporated $Cu(In,Ga)Se_2$ thin films

The morphology of the $Cu(In,Ga)Se_2$ samples is very uniform, compact and textured, composed of small particles (see figures 9a - 9c). Figure 9d shows the cross-section SEM image and a film thickness ~ 3.5 μm, also notice the details of the textured surface of the film, due to the high temperature processing.

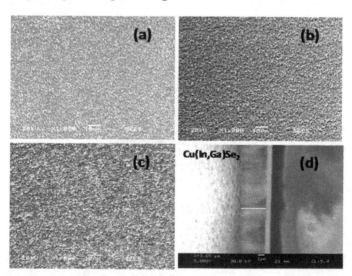

Fig. 9. SEM micrographs of co-evaporated $Cu(In,Ga)Se_2$ thin films (a - c) and (d) cross section image

The XRD patterns of the films show sharp and well defined peaks, indicating a very good crystallization, the films appear to grow with a strong (112) orientation (see figure 10) and with grain sizes ~ 1 μm. The expected shift of the (112) reflection compared to that of the $CuInSe_2$ is also observed, which is consistent with a film stoichiometry of $CuIn_{0.7}Ga_{0.3}Se_2$ (JCPDS 35-1102).

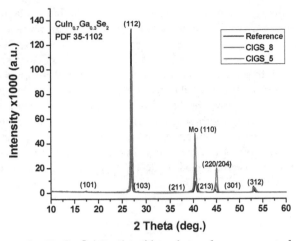

Fig. 10. XRD pattern for Cu(In,Ga)Se₂ thin films thermal co-evaporated

4. CdTe thin films by Close Spaced Vapor Transport (CSVT)

CdTe is a compound semiconductor of II-VI type that has a cubic zincblende (sphalerite) structure with a lattice constant of 6.481 A°. CdTe at room temperature has a direct band gap of 1.5 eV with a temperature coefficient of 2.3–5.4 x10^{-4} eV/K. This band gap is an ideal match to the solar spectrum for a photovoltaic absorber. Similarly to the Cu(In,Ga)Se₂, the absorption coefficient is large (around 5x10^4 cm^{-1}) at photon energies of 1.8 eV or higher (Birkmire R. and Eser E., 1997). Up to date the highest conversion efficiency achieved for CdTe solar cells is 16.5% (Wu X. et al., 2001). CdTe solar cells are p-n heterojunction devices in which a thin film of CdS forms the n-type window layer. As in the case of Cu(In,Ga)Se₂-based devices the depletion field is mostly in the CdTe. There are several deposition techniques to grow the CdTe like, physical vapor deposition, vapor transport deposition, close spaced sublimation, sputter deposition and electrodeposition (McCandless Brian E. and Sites James R., 2003). In this case, the close spaced sublimation has been selected to prepare the CdTe films for solar cell applications.

The sublimation technique for the deposition of semiconducting thin films of the II-VI group, particularly CdTe, has proven to be effective to obtain polycrystalline materials with very good optical and electrical properties. There are several steps that involve the formation of the deposited materials, these are listed as follows: 1) synthesis of the material to be deposited through the phase transition from solid or liquid to the vapor phase 2) vapor transport between the evaporation source and the substrate, where the material will be deposited in the form of thin film, and 3) vapor and gas condensation on the substrate, followed by the nucleation and grow of the films. In general, and particularly in our CdTe - case, the vapor transport is regulated by a diffusion gas model. This technique has several advantages over others because is inexpensive, has high growth rates, and it can be scaled up to large areas for mass production. The Close Spaced Vapor Transport technique, named as "CSVT", is a variant of the sublimation technique, it uses two graphite blocks, where independent high electrical currents flow and due to the dissipation effect of the electrical energy by Joule's heat makes the temperature in each graphite block to rise. One of the graphite blocks is named the source block and the other is the substrate block. Figure 11 shows the block diagram of the CSVT system used to prepare the CdTe thin films. Between the source graphite block "A" and the substrate graphite block "B" is located the graphite boat that contains the material to be sublimated, and on top of this boat the substrate is located, in a very close proximity or close spaced. The material growth is carried out under the presence of an inert atmosphere like argon, nitrogen, etc. The growth rate of the material to be deposited can be controlled by controlling the pressure and gas flow rate. Also this inert gas can be mixed with a reactive gas like oxygen, which benefits the growth of CdTe with the characteristic p-type conductivity. The deposition parameters for this technique are: a) T_s: temperature of the source, b) T_{sub}: substrate temperature, it has to be lower than the T_s in order to avoid the re-sublimation of the material, c) d_{s-sub}: distance between the material to be deposited and the substrate and d) P_g: pressure of the inert gas inside the chamber.

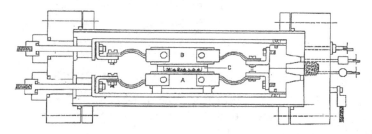

Fig. 11. Schematic diagram of a CSVT system

For the processing of CdTe thin film solar cells, it is necessary to use a *superstrate* structure, so that the CdS is deposited on SnO_2:F, in such a way that the growth process allows the film to be deposited over the whole surface, becoming a surface free of holes and caverns without empty spaces among the grains, and with a uniform grain size distribution. It is also required that the CdS layer matches well with the CdTe host, thus favoring a good growing kinetics for CdTe, as well as the formation of the CdS_xTe_{1-x} ternary compound in the interface due to the diffusion of S from CdS to CdTe. The high-efficiency CdTe solar cells to date have essentially the same *superstrate* structure. The superstrate structure is composed of a sodalime glass substrate, coated with a SnO_2:F; a transparent conductor oxide as the front contact, then a CdS layer is chemically bath deposited, followed by the deposition of a CdTe layer and finally the deposition of two layers of Cu and Au to form the back contact to complete the CdS/CdTe device. In order to achieve solar cells with high conversion efficiencies, the physical and chemical properties of each layer must be optimized (Morales-Acevedo A., 2006). The deposition of CdTe was performed by using CdTe powder 99.999% purity. The deposition atmosphere was a mixture of Ar and O_2, with equal partial pressures of O_2 and Ar. In all cases the total pressure was 0.1 Torr. Prior to all depositions the system was pumped to 8×10^{-6} Torr as the base pressure. In the CSVT-HW (hot wall) deposition, the separation between source and substrate was about 1 mm. The deposition time was 3 min for all the samples deposited with substrate and source temperatures of 550 °C and 650 °C, respectively. Under these conditions, CdTe layers of 2 – 4 μm were obtained. The CdTe thin films were also thermally treated with $CdCl_2$. As discussed before, a very important treatment independently of the deposition technique for both CdS and CdTe layers is a thermal annealing after the deposition of $CdCl_2$ on top of the CdTe layer. If the $CdCl_2$

treatment is not performed, the short circuit current density and the efficiency of the solar cell are very low. This treatment consists in depositing 300–400 nm of CdCl$_2$ on top of CdTe with a subsequent annealing at 400 °C during 15–20 min in air, or in an inert gas atmosphere like Ar. During this process the small CdTe grains are put in vapor phase and re-crystallize, giving a better-organized CdTe matrix. The presence of Cl$_2$ could favor the CdTe grain growth by means of a local vapor phase transport. In this way the small grains disappear and the CdS/CdTe interface is reorganized.

5. Processing of Cu(In,Ga)Se$_2$ and CdTe thin films into solar cells

Cu(In,Ga)Se$_2$ and CdTe PV devices are obtained by forming p-n heterojunctions with thin films of CdS. In this type of structure, n-type CdS, which has a band gap of 2.4 eV, not only forms the p-n junction with p-type CuInSe$_2$ or p-type CdTe but also serves as a window layer that lets light through with relatively small absorption. Also, because the carrier density in CdS is much larger than in CuInSe$_2$ or CdTe, the depletion field is entirely in the absorber film where electron-hole pairs are generated (Birkmire and Eser, 1997). After solar cell completion the photovoltaic parameters like I$_{sc}$, V$_{oc}$, FF and conversion efficiency were tested by doing the I-V characterization for the two structures; CdTe and Cu(In,Ga)Se$_2$. All the parameters were measured under AM1.5 illumination.

5.1 Cu(In,Ga)Se$_2$/CdS thin film solar cells
The substrate structure of a Cu(In,Ga)Se$_2$ thin film based solar cell is composed of a soda lime glass substrate, coated with a sputtered ~ 0.7 – 1 μm Mo layer as the back contact. After the thermal co-evaporation of Cu(InGa)Se$_2$ deposition, the junction is formed by chemically bath depositing the CdS with thickness ~ 30 - 50 nm. Then a high-resistance (HR) ZnO layer and a doped high-conductivity ZnO:Al layer are subsequently deposited, usually using the sputtering technique. Finally, the deposition of a current-collecting grid of Ni/Al completes the device as shown in figure 12. The total cell area is defined by removing the layers on top of the Mo outside the cell area by mechanical scribing.

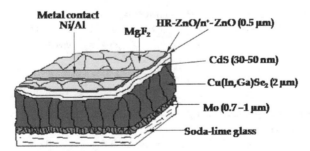

Fig. 12. Schematic configuration of a typical Cu(In,Ga)Se$_2$ thin film solar cell

5.1.1 Discussion on the Cu(In,Ga)Se$_2$ thin film based solar cells results
Two Cu(In,Ga)Se$_2$ samples were used to be processed into solar cell devices: sample JS17 had a CdS layer prepared with a recipe based on CdCl$_2$ and sample JS18 with a recipe based on CdSO$_4$ as the Cd source. The J-V parameters for devices JS17 are: area = 0.47 cm^2, V$_{oc}$ = 536 mV, J$_{sc}$ = 31.70 mA/cm^2, fill factor = 64.0 %, and η = 10.9 % (see figure 13) and for JS18 are: area

= 0.47 cm², V_{oc} = 558 mV, J_{sc} = 29.90 mA/cm², fill factor = 63.1 %, and η = 10.5 % (see figure 14). From these results, we can see that sample JS17 shows a conversion efficiency a little bit higher than JS18, this is due to the different recipe used to prepare the CdS layer as it was mentioned before. This was the only difference between the two devices, everything else was the same. From these figures, a low V_{oc} is observed, but we should expect to have a higher V_{oc} value, compared to the Ga content. The roll-over in forward bias could be indicative of a low sodium content in the Cu(In,Ga)Se₂ films. Also, the low current collection, observed for the Cu(In,Ga)Se₂ thin film devices, may be due to incomplete processing of the absorber layer. Improvements in device performance are expected with optimization of absorber processing.

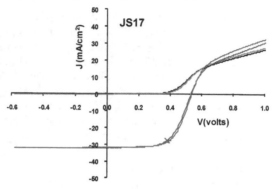

Fig. 13. J-V curves for the best Cu(In,Ga)Se₂ thin film device prepared with a CdS bath solution based on CdCl₂

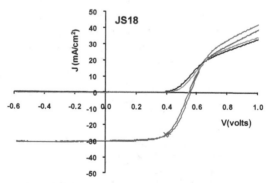

Fig. 14. J-V curves for the best Cu(In,Ga)Se₂ thin film device prepared with a CdS bath solution based on CdSO₄

5.2 CdTe/CdS thin film solar cells

The typical superstrate structure of a hetero-junction CdTe/CdS solar cell is composed of a soda lime glass substrate, coated with a sputtered transparent conducting oxide (TCO) to the visible radiation, which acts as the front contact, then a CdS layer with a thickness ~ 120 nm is chemically bath deposited, followed by the deposition of the absorber CdTe layer by close spaced vapor transport technique and finally the CdS/CdTe device is completed by depositing the ohmic back contact on top of the CdTe layer, see figure 15. For the back contact, two layers of Cu and Au (2nm and 350 nm, respectively) were evaporated, with an area of 0.08 cm², onto the CdTe and annealed at 180 °C in Ar. The front contact was taken from the conducting glass substrate (0.5 μm thick SnO₂:F/ glass with sheet resistivity of 10 Ω/█).

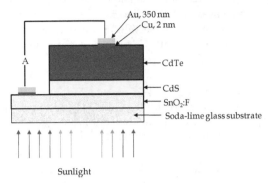

Fig. 15. Schematic configuration of a typical CdTe based solar cell

5.2.1 Variation of the S/Cd ratio in the solution for deposition of CdS by chemical bath and its effect on the efficiency of CdS/CdTe solar cells

The variation of the S/Cd ratio in the solution used in the preparation of the CdS films modifies the morphology, the deposition rate, the crystal grain size, the resistivity and the optical transmittance of these films and have an influence upon the structural and electrical properties of the CdTe layer itself, in addition to modifications of the CdS–CdTe interface. Hence, our study shows the influence of the S/Cd ratio in the solution for CdS thin films prepared by chemical bath upon the characteristics of CdS/CdTe solar cells with a superstrate structure (Vigil-Galán, et al., 2005).

The concentrations of NH₃, NH₄Cl and CdCl₂ were kept constant in every experiment, but the thiourea [CS(NH₂)₂] concentration was varied in order to obtain different S/Cd relations (R_{tc}) in the solution. All the films were grown on SnO₂:F conducting glasses (10 ohm-cm) at 75 °C. Deposition times were also varied, according to our previous knowledge of the growth kinetics (Vigil O. et al., 2001), with the purpose of obtaining films with similar thickness in all cases. The selected thiourea concentrations and deposition times for each S/Cd relation are shown in table 2.

S/Cd ratio R_{tc}	Thiourea concentration in the bath (mol/l)	Deposition time (min)
1	2.4×10^{-3}	120
2.5	6×10^{-2}	100
5	1.2×10^{-2}	120
10	2.4×10^{-2}	120

Table 2. Thiourea concentration and deposition time for each S/Cd relation

Solar cells were prepared by depositing CdTe thin films on the SnO₂:F/CBD-CdS substrates by CSVT-HW. The atmosphere used during the CdTe was a mixture of Ar and O₂, with an O₂ partial pressure of 50%. In all cases, the total pressure was 0.1 Torr. Prior to deposition the system was pumped to 8×10^{-6} Torr as the base pressure. CSVT-HW deposition of CdTe was done by placing a CdTe graphite source block in close proximity (1 mm) to the substrate block. The deposition time was 3 min for all the samples deposited with substrate and source temperatures of 550 °C and 650 °C, respectively. Under these conditions, CdTe layers of approximately 3.5 μm were obtained. The CdTe thin films were coated with a 200 nm $CdCl_2$ layer and then annealed at 400 °C for 30 min in air. For the back contact, two layers of Cu and Au (2 nm and 350 nm, respectively) were evaporated, with an area of 0.08 cm^2, on the CdTe film and annealed at 180 °C in Ar. The growth conditions of CdTe were maintained constant for all solar cells.

5.2.2 Discussion on CdTe thin film solar cells results

Figure 16 shows the set of I–V characteristics for CdS/CdTe solar cells made with the same R_{tc} (S/Cd ratio = 5). According to our experimental conditions, the solar cells made with the same technological process have similar characteristics.

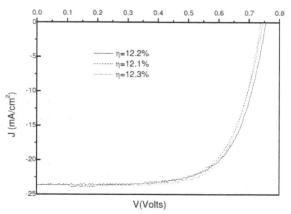

Fig. 16. J –V characteristics of three CdS/CdTe solar cells made with CdS layers grown with R_{tc} = 5 during the CBD-CdS growth process

The I–V characteristics of CdS/CdTe solar cells under AM1.5 illumination (normalized to 100 mW cm^{-2}) as a function of R_{tc} are shown in figure 17. In table 3, the average shunt (R_p) and series (R_s) resistances, the short circuit current density (J_{sc}), the open circuit voltage (V_{oc}), the fill factor (FF) and the efficiency (η) of solar cells prepared with different R_{tc} are reported. The averages were taken from four samples for each R_{tc}. As can be seen in table 3, η increases with R_{tc} up to R_{tc} = 5 and drops for R_{tc} = 10.

S/Cd ratio R_{tc}	R_s (ohm-cm^2)	R_p (ohm-cm^2)	J_{sc} (mA/cm^2)	V_{oc} (mV)	FF (%)	η (%)
1	6.8	318	20.8	617	55.2	7.1
2.5	5.4	800	21.8	690	55.5	8.3
5	2.9	787	23.8	740	70.5	12.3
10	5.9	135	22.7	435	52	5.4

Table 3. Photovoltaic parameter results for CdS/CdTe solar cells with different S/Cd ratio (R_{tc}) in the CdS bath

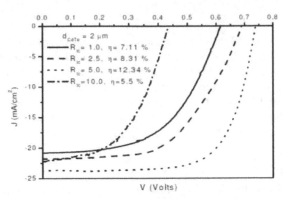

Fig. 17. Typical J – V characteristics of CdS/CdTe solar cells under illumination at 100 mW cm^{-2}, with R_{tc} as a parameter

There are several factors directly or indirectly influencing the cell behaviour, in particular the amount of S in the CBD CdS layers may influence the formation of the CdS$_{1-x}$Te$_x$ ternary compound at the CdS–CdTe interface. CdTe films grown at high temperatures, such as those produced by CSVT, produce a sulfur enriched region due to S diffusion. The amount of S penetrating the bulk of CdTe from the grain boundary must be dictated by the bulk diffusion coefficient of S in CdTe and of course by the amount of S available in the CdS films. The re-crystallization of CdTe could be affected by the morphological properties of the CdS layers grown with different S/Cd ratios. These facts have been studied by Lane (Lane D. W. et al., 2003) and Cousins (Cousins M. A. et al. 2003). From this point of view the formation of CdS$_{1-x}$Te$_x$ may be favored when the R_{tc} is increased in the bath solution. This ternary compound at the interface may cause a lower lattice mismatch between CdS and CdTe, and therefore a lower density of states at the CdTe interface region will be obtained, causing a lower value for the dark saturation current density J_0. The resistivity of the CdS and CdTe layers and their variation under illumination also change the characteristics of the cell under dark and illumination conditions. In other words, a better photoconductivity implies smaller resistivity values under illumination, with the possible improvement of the solar cell properties. In addition, optical transmittance, thickness and morphological measurements of the CBD-CdS films showed the following characteristics when increasing R_{tc}: i) band gap values are observed to increase (from 2.45 eV to 2.52 eV when changing R_{tc} from 1 to 10), ii) grain sizes become smaller (from 55.4 nm to 47.2 nm when S/Cd = 1 and 10, respectively) and iii) the average optical transmission above threshold increases from 68% to 72% when R_{tc} is increased from 1 to 10. Higher band-gap values of the window material improve the short circuit current density of the solar cells. Thin films with smaller grain sizes show fewer pinholes with a positive effect on the open circuit voltage and fill factor. In this regard, the properties of the CdS layers are correlated with the kinetic of the deposition process when the concentration of thiourea is changed. For instance, for high thiourea concentration, the reaction rate becomes large enough to promote a quick CdS precipitation which leads to the formation of agglomerates in the solution rather than nucleation on the substrate surface, while for low thiourea concentration a very slow growth process can be expected, leading to a thinner but more homogeneous layer.

6. Conclusions

We have found that CBD-CdS thin films grown under different conditions, like monolayers or bi-layers, using a standard bath configuration or a modified configuration, the principle for the deposition process is the same: a common precipitation reaction. Depending of the regime we decide to choose, we must perform an optimization of the deposition parameters in order to get the CdS film with the best physical and chemical properties. The quality of the CdS window partner and the absorber material like CdTe and $Cu(In,Ga)Se_2$ will have a great impact on the conversion efficiencies when applied into thin film solar cells.

7. Acknowledgements

The authors would like to thank to Bill Shafarman from University of Delaware for device processing and characterization. This work was partially supported by CONACYT, grant 47587, ICyT-DF, grant PICS08-54 and PROMEP, grant 103.5/10/4959.

8. References

Birkmire R. and Eser E. (1997), Annu. Rev. Mater. Sci., 27:625–53.

Boyd D, Thompson D, Kirk-Othmer, (1980), Encyclopaedia of Chemical Technology, Vol. 11, 3rd Edition, 807–880, John Wiley.

Contreras-Puente G., Tufino-Velazquez M., Calixto M. Estela, Jimenez-Escamilla M., Vigil-Galan O., Arias-Carbajal A., Morales-Acevedo A., Aguilar-Hernandez J., Sastre-Hernandez J., Arellano-Guerrero F.N. (2006), Photovoltaic Energy Conversion, Conference Record of the 2006 IEEE 4th World Conference.

Cousins M. A., Lane D. W., and Rogers K. D. (2003). *Thin Solid Films* 431–432, 78.

Lane D. W., Painter J. D., Cousins M. A., Conibeer G. L., and Rogers K. D. (2003). Thin Solid Films 431–432, 73.

McCandless Brian E. and Sites James R., (2003), Cadmium Telluride Solar Cells, in: Handbook of Photovoltaic Science and Engineering, Luque A. and Hegedus S., pp. 567 – 616, John Wiley & Sons, Ltd ISBN: 0-471-49196-9, West Sussex, England.

Mitchell K. et al., (1990), IEEE Trans. Electron. Devices 37, 410–417.

Morales-Acevedo A. (2006). Solar Energy 80, 675.

Repins I. et al., (2008), Prog. in Photov: Research and Applications 16, 235.

Rocheleau R, Meakin J, Birkmire R, (1987), Proc. 19th IEEE Photovoltaic Specialist Conf., 972–976.

Romeo N., Bosio A., Canevari V. and Podestà A. (2004), Solar Energy, Volume 77, Issue 6, Pages 795-801.

Shafarman W. N. and Stolt L. (2003), Cu(InGa)Se2 Solar Cells, in: Handbook of Photovoltaic Science and Engineering, Luque A. and Hegedus S., pp. 567 – 616, John Wiley & Sons, Ltd ISBN: 0-471-49196-9, West Sussex, England.

Schon, J. H., Schenker, O. and Batlogg, B. (2001), Thin Solid Films 385, p.271.

Vigil-Galan O., Arias-Carbajal A., Mendoza-Perez R., Santana-Rodríguez G., Sastre-Hernandez J., Alonso J. C., Moreno-Garcia E., Contreras-Puente G. and Morales-Acevedo A. (2005), Semiconductor Science and Technology 20, 819.

Vigil O., Arias-Carbajal A., Cruz F., Contreras-Puente G., and Zelaya-Angel O. (2001). Mats. Res. Bull. 36, 521.

Wu X. et al., (2001), Proc. 17th European Photovoltaic Solar Energy Conf. 995.

Influence of Post-Deposition Thermal Treatment on the Opto-Electronic Properties of Materials for CdTe/CdS Solar Cells

Nicola Armani[1], Samantha Mazzamuto[2] and Lidice Vaillant-Roca[3]

[1]IMEM-CNR, Parma
[2]Thifilab, University of Parma, Parma
[3]Lab. of Semicond. and Solar Cells, Inst. of Sci. and Tech. of Mat.,
Univ. of Havana, La Habana
[1,2]Italy
[3]Cuba

1. Introduction

Thin film solar cells based on polycrystalline Cadmium Telluride (CdTe) reached a record efficiencies of 16.5% (Wu et al. 2001a) for laboratory scale device and of 10.9% for terrestrial module (Cunningham, 2000) about ten years ago. CdTe-based modules production companies have already made the transition from pilot scale development to large manufacturing facilities. This success is attributable to the peculiar physical properties of CdTe which make it ideal for converting solar energy into useful electricity at an efficiency level comparable to silicon, but by consuming only about 1% of the semiconductor material required by Si solar cells. Because of the easy up-scaling to an industrial production as well as the low cost achieved in the recent years by the manufacturers, the CdTe technology has carved out a remarkable part of the photovoltaic market. Up to now two companies (Antec Solar and First Solar) have a noticeable production of CdTe based modules, which are assessed as the best efficiency/cost ratio among all the photovoltaic technologies.

Since the record efficiency of such type solar cells is considerably lower than the theoretical limit of 28-30% (Sze, 1981), the performance of the modules, through new advances in fundamental material science and engineering, and device processing can be improved. Further studies are required to reveal the physical processes determining the photoelectric characteristics and the factors limiting the efficiency of the devices.

The turning point for obtaining the aforementioned high efficiency values was the application of a Cl-based thermal treatment to the structures after depositing the CdTe layer (Birkmire & Meyers, 1994; McCandless & Birkmire, 1991). The device performance improvement is due to a combined beneficial effect on the materials properties and on the p-n junction characteristics. CdTe grain size increase (Enriquez & Mathew, 2004; Luschitz et al., 2009), texture properties variations (Moutinho et al., 1998), grain boundary passivation, as well as strain reduction due to S diffusion from CdS to the CdTe layer and recrystallization mechanism (McCandless et al., 1997) are the common observed effects.

In the conventional treatment, based on a solution method, the as-deposited CdTe is coated by a $CdCl_2$ layer and then annealed in air or inert gas atmosphere at high temperature. Afterwards, an etching is usually made to remove some $CdCl_2$ residuals and oxides and to leave a Te-rich CdTe surface ready for the back contact deposition. This etching is usually carried out with a Br-methanol solution or by using a mixture of HNO_3 and HPO_3. Alternative methodologies avoiding the use of solutions have been developed: the CdTe films are heated in presence of $CdCl_2$ vapor or a mixture made by $CdCl_2$ and Cl_2 vapor, or HCl (Paulson & Dutta 2000). Vapor based treatments reduce processing time since combining the exposure to $CdCl_2$ and annealing into one step.

All these post-deposition treatments have been demonstrated to strongly affect the morphological, structural and opto-electronic properties of the structures. The changes induced by the chlorine based treatments depend on how the CdTe and CdS were deposited. For example, in CdTe films having an initial sub micrometer grain size, it promotes a recrystallization mechanism, followed by an increase of the grains. This recrystallization process takes place in all CdTe films having specific initial physical properties, and does not depend on the deposition method used to grow the films. Recrystallization together to grain size increase has been observed in CdTe films deposited by Closed Space Sublimation (CSS), Physical Vapor Deposition (PVD) or Radio Frequency Sputtering. The chlorine based treatment may or may not induce recrystallization of the CdTe films, depending on the initial stress state of the material, and the type and conditions of the treatment. For this reason, the recrystallization process wasn't observed in CSS samples which are deposited at higher temperatures and have an initial large grain size, while, for example CdTe films deposited by Sputtering that are characterized by small grains lower than 1µm in size, an increase up to one order of magnitude was obtained (Moutinho et al., 1998, 1999). The driving force for the recrystallization process is the lattice-strain energy at the times and temperatures used in the treatment.

Changes in structural properties and preferred orientation are also observed. The untreated CdTe material usually grows in the cubic zincblende structure, with a preferential orientation along the (111) direction. Depending on the deposition method, these texture properties can be lost, in place of a completely disoriented material. The Cl-based annealing induces a lost of the preferential orientation as demonstrated by literature X-Ray Diffraction (XRD) works explaining in terms of σ value calculation (Moutinho et al. 1998, 1999). However, this treatment is important even in films that do not recrystallize because it decreases the density of deep levels inside the bandgap and changes the defect structure, resulting in better devices.

Maybe the crucial effect of the treatment is related to the p-n junction characteristics. This treatment promotes interdiffusion between CdTe and CdS, resulting in the formation of CdTeS alloys at the CdTe–CdS interface. The $CdTe_{1-x}S_x$ and $CdS_{1-y}Te_y$ alloys form via diffusion across the interface during CdTe deposition and post-deposition treatments and affect photocurrent and junction behavior (McCandless & Sites, 2003).

Formation of the $CdS_{1-y}Te_y$ alloy on the S-rich side of the junction reduces the band gap and increases absorption which reduces photocurrent in the 500–600 nm range. Formation of the $CdTe_{1-x}S_x$ alloy on the Te-rich side of the junction reduces the absorber layer bandgap, due to the relatively large optical bowing parameter of the CdTe–CdS alloy system.

Despite the promising results, the transfer to an industrial production of the commonly adopted $CdCl_2$ based annealing may increase the number of process steps and consequently the device final cost (Ferekides et al., 2000). Since $CdCl_2$ has a quite low evaporation temperature (about 500°C in air), it cannot be stored in a large quantity, since it is dangerous because it can release Cd in the environment in case of fire. Secondly, $CdCl_2$ is soluble in water and, as a consequence, severe security measures must be taken to preserve environmental pollution and health damage. Another drawback is related to the use of chemical etchings, such as HNO_3 and HPO_3 or Br-Methanol solution, implying that a proper disposal of the used reagents has to be adopted since the workers safety in the factory must be guaranteed. In order to overcome the aforementioned drawbacks, we substituted the $CdCl_2$ based process with an alternative, completely dry CdTe post-deposition thermal treatment, based on the use of a mixture of Ar and a gas belonging to the Freon family and containing chlorine, such as difluorochloromethane (HCF_2Cl)(Bosio et al., 2006, Romeo N. et al., 2005). This gas is stable and inert at room temperature and it has not any toxic action. Moreover, the post-treatment chemical etching procedures have been eliminated by substituting them with a simple vacuum annealing.

The only drawback in using a Freon gas could be that it is an ozone depleting agent, but, in an industrial production, it can be completely recovered and reused in a closed loop. In this paper, it will be demonstrated how the CdTe treatment in a Freon atmosphere works as well as the treatment carried out in presence of $CdCl_2$.

This method was successfully applied to Closed Space Sublimation (CSS) CdS/CdTe solar cells, by obtaining high-efficiency up to 15% devices (Romeo N. et al., 2007). This original approach may produce modifications on the material properties, different than the usual $CdCl_2$-based annealing. For this reason, in this work, the efforts are focused on the investigation of the peculiar effects of the treatment conditions on the morphology, structural and luminescence properties of CdTe thin films deposited by CSS on Soda-Lime glass/TCO/CdS. All the samples were deposited by keeping unmodified the growth parameters (temperatures and layer thicknesses), in order to submit as identical as possible materials to the annealing. Only the HCF_2Cl partial pressure and the Ar total pressure in the annealing chamber have been varied.

The aim of the present work is to correlate the effect of this new, all dry post-deposition treatment, on the sub-micrometric electro-optical properties of the CSS deposited CdTe films, with the effect on the device performances. Large area SEM-cathodoluminescence (CL) analyses have allowed us to observe an increase of the overall luminescence efficiency and in particular a clear correlation between the defects related CL band and the HCF_2Cl partial pressure in the annealing atmosphere. By the high spatial (lateral as well as in-depth) resolution of CL, a sub-micrometric investigation of the single grain radiative recombination activity and of the segregation of the atomic species, coming from the Freon gas, into grain boundary has been performed.

The HCF_2Cl partial pressure has been changed from 20 to 50 mbar, in order to discriminate the Freon gas effect from the others annealing parameters. A clear correlation between the CL band intensities and the HCF_2Cl partial pressure has been found and a dependence on the lateral luminescence distribution has been observed.

The results obtained from the material analyses have been correlated to the performances of the solar cells processed starting from the glass/ITO/ZnO/CdS/CdTe structures studied. Electrical measurements in dark and under illumination were carried out, in order to determine the characteristic photovoltaic parameters of the cell and to investigate the transport processes that take place at the junction. In particular the device short circuit current density (J_{SC}), open circuit voltage (V_{OC}) fill factor (ff) and efficiency (η) have been measured as a function of the HCF_2Cl partial pressure. The most efficient device obtained by this procedure, corresponding to 40 mbar HCF_2Cl partial pressure in the 400mbar Ar total pressure, has η=14.8%, J_{SC}=26.2mA/cm², V_{OC} = 820mV and ff=0.69.

The solar cells were then submitted to an etching procedure in a Br–methanol mixture at 10% to eliminate the back contacts and part of the CdTe material in some portion of the specimens. On the beveled surface, CL analyses have been performed again in order to extract information as close as possible to the CdTe/CdS interface and to compare the results to the depth-dependent CL analyses.

Finally, a model of the electronic levels present in the CdTe bandgap before and after the HCF_2Cl treatment has been proposed as well as a model of the interface region modifications due to the annealing.

2. Materials growth and devices preparation

CdTe is a II-VI semiconductor with a direct energy-gap of 1.45eV at room temperature that, combined with the very high absorption coefficient, 10^4-10^5 cm^{-1} in the visible light range, makes it one of the ideal materials for photovoltaic conversion, because a layer thickness of a few micrometers is sufficient to absorb 90% of incident photons. For thin film solar cells is required a p-type material, which is part of the p-CdTe/n-CdS heterojunction. The electrical properties control was easily developed for single-crystal CdTe, grown from the melt or vapor, at high temperature (above 1000°C), by introducing doping elements during growth. On the contrary, in polycrystalline CdTe, where grain boundaries are present, all metallic dopants tend to diffuse along the grain boundaries, making the doping unable to modify the electrical properties and producing shunts in the device.

CdTe solar cell is composed by four parts (Fig. 1) deposited on a substrate like Soda-Lime Glass (SLG):

1. The *Front Contact* is composed by a Transparent Conducting Oxide (TCO) that is a doped metallic oxide like In_2O_3:Sn (ITO)(Romeo N. et al., 2010) , ZnO:Al (AZO)(Perrenoud et al., 2011), $CdSnO_4$ (CTO)(Wu, 2004), SnO_2:F (FTO) (Ferekides et al., 2000), etc.; and a very thin layer of a resistive metal oxide like SnO_2 (Ferekides et al., 2000), ZnO (Perrenoud et al., 2011; Romeo N. et al. 2010), Zn_2SnO_4 (Wu et al. 2001b). The role of the latter film is to hinder the diffusion of contaminant species from TCO and SLG toward the upper layers of the cell such as the window layer (CdS) or the absorber one (CdTe). Moreover it separates TCO and CdS in order to limit the effects of pinholes that could be present in CdS film.

   In our work, TCO is made by 400nm thick ITO film and 300nm thick ZnO both of them deposited by sputtering. ITO showed a sheet resistance of about 5Ω/cm², while the resistivity of ZnO was on the order of 10^3 Ω cm.

2. The *Window Layer* is usually an n-type semiconductor; Cadmium Sulphide (CdS) is the most suitable material for CdTe-based solar cells, thanks to its large bandgap (2.4eV at room temperature) and because it grows with n-type conductivity without the introduction of any dopants. Here, CdS film was deposited by reactive RF sputtering in presence of Ar+10%CHF_3 flux. Its nominal thickness was 80nm.

3. The *Absorber Layer* is a 6-10μm thick film. The deposition techniques and the treatment on CdTe will be explained deeply later.

4. The *Back Contact* is composed by a buffer layer and a Mo or W film. The utility of the buffer layer is to form a low resistive and ohmic contact on CdTe.

The cell is completed by a scribing made on the edge of all the cells in order to electrically separate the front contact from the back one.

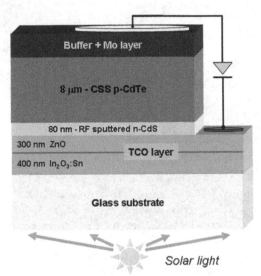

Fig. 1. Schematic representation of the CdS/CdTe solar cell heterostructure. The layers succession and thicknesses are the ones used in the present work.

2.1 CSS Growth of CdTe layers

CdTe thin films have been deposited by several deposition techniques such as High Vacuum Evaporation (HVE)(Romeo A. et al., 2000), Electro-Deposition (ED)(Josell et al., 2009; Kosyachenko et al., 2006; Levy-Clement, 2008; Lincot, 2005), Chemical Vapour Deposition (CVD)(Yi & Liou, 1995), Metal-Organic Chemical Vapor Deposition (MOCVD)(Barrioz, 2010; Hartley, 2001; Zoppi, 2006), Spray Pyrolysis (Schultz et al., 1997), Screen Printing (Yoshida, 1992 & 1995) Sputtering (Compaan et al., 1993; Hernández-Contreras et al., 2002; Plotnikov et al., 2011) and Close Spaced Sublimation (CSS)(Chu et al., 1991; Romeo N. et al., 2004; Wu, 2004).

Among these techniques, CdTe deposited by CSS allowed to obtain best results for solar cells (world record photovoltaic solar energy conversion ~16.5%; Wu, 2004).

CSS is a physical technique based on a high temperature process. The apparatus is showed in Fig. 2 and it is composed by a vacuum chamber inside which the substrate and the source are placed at a distance of few millimeters (2-7mm). The difference in temperature between the substrate and the source is kept around 50-150°C. Deposition takes place in presence of an inert gas (Ar) or a reactive one (O_2, etc.) with a total pressure of about 1-100mbar. The gas creates a counter-pressure which reduces re-evaporation from the substrate and forces the atoms from the source to be scattered many times by the gas atoms before arriving to the substrate, so that the material to be deposited acts like it has a higher dissociation temperature and higher temperature respect to sublimation under vacuum are necessary.

CSS allows to obtain CdTe film with a very high crystalline quality and grains of about one order of magnitude larger (~10μm) than films deposited by other deposition techniques (Sputtering, HVE, etc.) and, for this reason, with a low lattice defect density (Romeo A. et al., 2009).

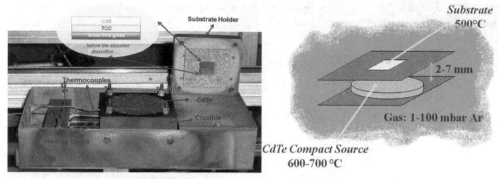

Fig. 2. Picture of the CSS setup used for growing the CdTe films studied (left); Detail of the growth region of the CSS chamber (right).

In our work, CdTe was deposited in 1mbar Ar atmosphere, keeping the substrate and source temperatures at 500°C and 600°C respectively. The CdTe thickness was 6-8 μm.

The high substrate temperature (~500°C) favors the formation of a mixed compound CdS_xTe_{1-x} at the interface between CdS and CdTe directly during CdTe deposition, as shown in the phase diagram (Lane et al., 2000). The mixed compound formation, by means of S diffusion toward CdTe and Te diffusion toward CdS, is advantageous in order to get high efficiency CdS/CdTe solar cells. In fact, its formation is required in order to minimize defect density at the interface acting as traps for majority carriers crossing the junction, caused by the lattice mismatch between CdS and CdTe that is about 10%.

2.2 HCF₂Cl post-deposition thermal treatment

The Cl-treatment on CdTe surface is a key point in order to rise the photocurrent and so the efficiency of the solar cell.

During Cl-treatment CdTe goes in vapor phase as explained by the following reaction (McCandless, 2001):

$$CdTe(s)+CdCl_2(s) \rightarrow 2Cd(g)+\tfrac{1}{2}Te_2(g)+Cl_2(g) \rightarrow CdCl_2(s)+CdTe(s), \qquad (1)$$

where s is the solid phase and g is the vapor phase.

After the treatment small grains disappear from CdTe surface and at the same time an increase in grain dimensions and an improvement in crystal organization can be observed. Also an improvement, in the crystal organization of the mixed compound CdS_xTe_{1-x}, at the junction, formed during CdTe deposition, can be observed.

Usually Cl-treatment is carried out by depositing on CdTe surface a $CdCl_2$ (thickness more than 100nm) film by evaporation (Potter et al., 2000; Romeo A. et al. 2000; Romeo N. et al. 1999) or by dipping CdTe in a $CdCl_2$-methanol solution (Cruz et al., 1999), then an annealing at ~380-420°C in an Ar atmosphere or in air is required and finally, an etching in Br-methanol or an annealing in vacuum is carried out in order to remove $CdCl_2$ residuals on CdTe surface. The main drawback of this treatment is that $CdCl_2$, being very hygroscopic, could be dangerous either for people and for the environment since it can release free Cd.

We have proposed a new treatment by substituting $CdCl_2$, or $CdCl_2$-methanol solution, and the following etching with a treatment at 400°C in a controlled atmosphere containing a gas belonging to the Freon® family which can free Cl at high temperature. This gas is very stable and inert at the room temperature; moreover, in the case of an industrial production, it can be re-used in a closed loop without releasing it in atmosphere.

We suppose that the following reaction happens at 400°C during the treatment (Romeo N. et al., 2006):

$$CdTe(s) + 2Cl_2(g) \rightarrow CdCl_2(g) + TeCl_2(g) \rightarrow CdTe(s) + 2Cl_2(g). \tag{2}$$

After that, an annealing is carried out at the same temperature of the treatment for few minutes in vacuum (10^{-5}mbar) in order to let $CdCl_2$ residuals re-evaporate and to obtain a clean CdTe surface ready for the back contact deposition.

In this work, the TCO/CdS/CdTe system is placed in an evacuable quartz ampoule. Before each run, the ampoule is evacuated with a turbo-molecular pump up to 10^{-6}mbar. As a source of Cl_2, a mixture of Ar+HCF$_2$Cl is used. The samples were prepared by changing the HCF$_2$Cl partial pressure. The first one was an untreated sample, while the other four ones were made by choosing four values of HCF$_2$Cl partial pressure that are 20, 30, 40, and 50 mbar and keeping the total pressure (Ar+HCF$_2$Cl) at 400 mbar. An additional specimen, annealed at 30mbar HCF$_2$Cl partial pressure, but with a larger total Ar+HCF$_2$Cl pressure of 800mbar, has been prepared, in order to study the effect of the total pressure on the recrystallization mechanisms. The Ar and the HCF$_2$Cl partial pressures were independently measured by two different capacitance vacuum gauges and monitored by a Varian Multi-Gauge. The quartz tube is put into an oven where a thermocouple is installed in order to control the furnace temperature, which is set at 400°C. The annealing time is 10min for all samples studied in this work. After the treatment, a vacuum for about 10min, keeping the temperature at 400°C, was made in order to remove some $CdCl_2$ residuals from the CdTe surface.

2.3 Back contact deposition and device processing

The cell is completed by back contact deposition. The formation of a ohmic and stable back contact with CdTe has always been one the most critical points in order to obtain high efficiency CdS/CdTe solar cells. Normally, CdTe is etched in order to get a Tellurium rich surface. After that, a Cu film (~2nm thick) is deposited in order to form a Cu$_x$Te compound that is a good non-rectifying contact for CdTe. This procedure has two disadvantages: chemical etching is not convenient because it is not scalable to an industrial level and it is polluting and the Cu thickness is too small to be controlled. In fact, if a thicker Cu film is deposited it could happen that Cu is free from the Cu$_x$Te formation and it could cause short circuits in the cell because it can segregate in grain boundaries.

In our work, back contact is composed by the deposition in sequence of three films. A 150-200nm thick As$_2$Te$_3$ film and a 10-20nm thick Cu film are deposited in sequence on CdTe surface by RF sputtering in Ar flux. When the deposition temperature of Cu is about 150-200°C, a substitution reaction occurs between Cu and As$_2$Te$_3$ whose final product material is Cu$_x$Te, mainly Cu$_{1.4}$Te is the most stable compound (Romeo N. et al, 2006; Wu et al. 2006; Zhou, 2007). Finally, a Mo layer is deposited on top of the cell by sputtering.

2.4 Etching procedures by a Br – methanol mixture

The possibility to perform depth-dependent CL analyses, by increasing the energy of the incident electrons of the SEM, allows us to correlate the results obtained on the isolated CdTe to an analysis of the electro-optical properties close to the CdTe/CdS interface region of a complete solar cell. To do this, it is necessary to overcome the problem that summing the back contact to the CdTe thickness, the main junction is situated around 10μm below the specimen surface. This thickness is 2 times higher than the maximum distance that the most energetic electrons in our SEM (40keV) penetrate in CdTe. In addition, it has to take into account that the back contact completely absorbs the light coming from the CdTe film, impeding any CL analyses.

In this work, a solution to this experimental difficulty has been proposed by etching the material to completely eliminate the back contact and the excess CdTe in some portion of the cells. In order to prevent the introduction of superficial defects that would affect the CL reliability, polishing methods were avoided. On the other hand, standard nitric–phosphoric acid chemical etching widely performed before metallization to improve contact formation, shows a strong preferential chemical reaction over the grain boundaries (Bätzner et al. 2001; Xiaonan et al. 1999). For these reasons, a Br–methanol mixture at 10% has been used, expecting to obtain a less selective interaction of the etching solution between the grains and its boundaries.

3. Experimental and set-up description

The methodological approach used in this work was based on the correlation between the study of HCF_2Cl treatment effect on CdTe material properties and the characterization of the photovoltaic cells parameters. There are not many works in literature that correlate the effects of CdTe post-deposition treatment and the relative changes in the electro-optical properties of CdTe with the performance of the photovoltaic device. Only recent studies (Consonni et al. 2006) on the behavior of Cl inside polycrystalline CdTe gave major results about the compensation mechanisms and the formation of complexes between native point defects (NPD) and impurities, already well established in the case of high quality single crystal CdTe (Stadler et al., 1995). The influence of post-deposition treatment on the CdTe/CdS interface region was crucial in the improvement of the device performances. The in-depth CdTe thin film properties, obtained by CL analyses, are then compared to results obtained on etched CdTe samples, treated in the same HCF_2Cl conditions. This allows us to verify the reliability of CL depth-resolution studies on polycrystalline materials and the effect of HCF_2Cl thermal treatment on the bulk CdTe properties approaching the CdTe/CdS interface.

3.1 Cathodoluminescence spectroscopy and mapping

CL is a powerful technique for studying the optical properties of semiconductors. It is based on the detection of the light emitted from a material excited by a highly energetic electron beam. The high-energy electron beam (acceleration voltage between 1-40kV), impinging on the sample surface, creates a large number of electron hole (e-h) pairs. After a thermalization process, the carriers reach the edges of the respective bands, conduction band (CB) in the case of electrons, valence band (VB) in the case of holes, and then diffuse. From the band edges, the electrons and holes can recombine, in the case of radiative recombination, the photons produce the CL signal. A more detailed description of the principles of the CL theory, in particular the fundamental of the generation and recombination mechanisms of the carriers can be found in the works of B. Yacobi and D. Holt (Yacobi & Holt, 1990) and references therein included.

CL is contemporary a microscopic and spectroscopic methodology with high spatial, lateral as well as in-depth, resolution and good spectral resolution when luminescence is detected at low temperature. These advantages are due to the use of a focused electron beam of a SEM as excitation source. In addition, this technique allows the contemporary acquisition of spectra of the intensity of the light collected as a function of wavelength and images (mono- and pan-chromatic) of the distribution of the light. The results can be acquired from regions of different area, from 1 to several hundreds of μm^2, depending on the magnification of the SEM and on the dimensions of the parabolic mirror used as light collector.

The lateral resolution in CL imaging can be roughly defined as the minimum detectable distance between two regions presenting different CL intensity. In the SEM-CL, the spatial imaging resolution depends mainly on the size of the recombination volume (generation volume broadened for the diffusion length) of e-h pairs inside the material, entailing also a dependence on the diffusion length (L) of generated carriers. A typical value of the lateral resolution of about 200nm can be reached as a lower limit in suitable working conditions for instance on III-V semiconducting quantum confined heterostructures (Merano et al., 2006).

The in-depth analysis is a CL peculiarity which allows us to investigate the samples at different depths by changing the energy of the primary electrons. The generation, as well as the recombination volume, increases in all the three dimensions by increasing the acceleration voltage. The depth, at which the maximum CL signal is created, increases also by increasing the beam energy (E_b). By this method, it is possible to investigate crystals or thick layers inhomogeneities along the growth direction. The large grain size of the CSS deposited CdTe, higher than 1 μm, allowed us to directly investigate the grain and the grain boundary recombination properties. This possibility is very useful to study a possible gettering mechanism or a passivation effect of the grain boundaries due to the annealing.

The post-deposition thermal treatment has an effect on the CdTe surface as well as on the bulk material, reasonably as far as the CdTe/CdS interface. For this reason, a complete characterization of the CdTe electro-optical properties and of the p-n junction recombination mechanisms, by using a bulk sensitive experimental technique, is necessary. The penetration depth of 200-300nm of the laser radiation used for PL analyses is a disadvantage that could be overcome by using the high energy electrons of an SEM for exciting CL. In addition, the possibility of increasing the CL generation/recombination volume by increasing the electron beam energy allows us a depth-dependent analysis. The CL analysis of 6-8μm thick CdTe thin films, as the active layers used in the fabrication of solar cells, has particular advantages: the maximum penetration depth of the exciting electrons of the SEM beam can reach 4.8μm by using 36 keV energy. This depth is higher than the few hundreds of nanometers probed by the commonly used Ar laser (514 nm) to excite PL. It is actually possible to perform an investigation of the CdTe bulk properties by CL in place of a near-surface PL analyses. The in-depth information, that is not available with other micro- and nano-scale optical techniques, is particularly useful for example in the characterization of heterostructures, doping profile, study of extended defects along the growth direction.

However, it is important to remark that the fundamental differences between CL and PL are the amount of e-h pairs generated and the dimensions and shape of the generation volumes. In the case of laser generation, each photon creates a single e-h pair whereas a high energetic electron can generate thousands of e-h pairs. With such a large number of e-h pairs generated, the excitation of all the radiative recombination channels inside the materials is possible.

The instrument used to collect the experimental data reviewed in this work is a Cambridge 360 Stereoscan SEM with a tungsten filament (resulting beam size on the sample surface typically ranging between a few microns and a few tens of nanometers), equipped with a Gatan MonoCL2 system (Fig. 3). The spectra, as well the panchromatic and monochromatic images, have been acquired using a dispersion system equipped with three diffraction gratings and a system of a Hamamatsu multi-alkali photomultiplier and a couple of liquid nitrogen cooled (Ge and InGaAs) solid state detectors. This experimental set up provides a spectral resolution of 2Å and a detectable 250-2200nm (0.6–4.9eV) wavelength range. By this configuration it is possible to cover a large part of the luminescence emissions of the III-V and II-VI compound semiconductors. In particular, all the possible transitions in CdTe can be detected: from the excitonic lines (around 1.59eV) down to the emissions involving mid-gap levels (0.8-0.9eV). Additionally, it is possible to change the temperature of the samples in the range 5-300K by a temperature controller interlocked with the sample-holder, thanks to a refrigerating system operating with liquid Nitrogen and liquid Helium.

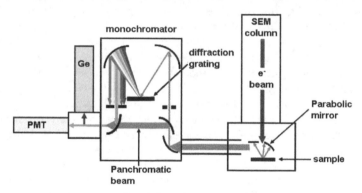

Fig. 3. Schematic representation, not in scale, of the CL experimental setup used in this work.

3.2 X-Ray diffraction
The setup used for acquiring XRD profiles was an X-Ray Diffractometer Thermo arl X'tra, vertical goniometer, theta-theta, operating in an angular range between -8° and 160°, equipped with an X-ray tube, Cu K-alpha and a solid state Si:Li detector. The angular range chosen, between 15° and 80°, assured the detection of all the contributions from the main Bragg diffractions of CdTe: (111), (220), (311), (400), (331), (422), (511).

3.3 Electrical characterization
Light J-V measurements were performed by an Oriel Corporation Solar Cells Test System model 81160, in order to measure the photovoltaic parameters such as the short-circuit current density (J_{SC}), open circuit voltage (V_{OC}), fill factor (ff) and conversion efficiency (η) of the solar cells.

Dark measurements were carried out by a Keithley 236 source system in order to measure the diode quality factor (A) of the cells as a function of the HCF_2Cl partial pressure during the CdTe treatment. A can be calculated from the diode equation in the dark:

$$A = \frac{qV}{kT} \cdot \frac{1}{\ln(\frac{J}{J_0} + 1)} \tag{3}$$

The measure of A gave some information about the transport mechanism at the junction. If the predominant transport mechanism at the junction is the diffusion then A≈1, while if the predominant mechanism is the recombination, the value of A increased and approached to 2. The dark conductivity as a function of the temperature (84-300K) and the activation energy were performed by using a Keithley 236 source measure unit. The temperature was set by a system DL4600 Bio-Rad Microscience Division. The samples, used for this measurement, were composed by 300nm thick ZnO, 7μm thick CdTe and the back contact. The first sample was a not treated one, while the other two samples were made by treating CdTe with respectively 30 and 40mbar HCF_2Cl partial pressure at 400°C for 10 minutes. The total pressure (Ar+Freon®) was set at 400mbar for all the two samples.

4. Results and discussion

All the CdTe thin films were deposited on SLG/ZnO substrate by CSS; the layer thickness was about 8μm. Complete solar cells have been realized by depositing ZnO, CdS and CdTe in the identical conditions and by adding the back contact, as described in paragraphs 2.1 and 2.3. The CdTe films as well as the complete devices were annealed in Ar+HCF_2Cl atmosphere (see for details paragraph 2.2), by increasing the HCF_2Cl partial pressure from 20mbar to 50mbar and keeping the temperature at 400°C for all samples. The annealing conditions used have been summarized in table 1.

Sample	HCF_2Cl partial pressure (mbar)	Ar+HCF_2Cl total pressure (mbar)	Annealing time (mins)
UT	-	-	-
F20	20	400	10
F30L	30	400	10
F30H	30	800	5
F40	40	400	10
F50	50	400	10

Table 1. Summary of the annealing conditions used to treat the samples studied in this work

4.1 Influence of annealing on the CdTe material properties
The XRD profiles of all the CdTe films were acquired in the angular range 5°<2q<80°, from this analysis can be deduced that the films have a zinc-blend structure with a preferential orientation along the (111) direction. In all the XRD patterns the peaks related to (220), (311), (400), (331), (422) and (511) reflections are also visible. In addition a peak at 22.77° attributed to the Te_2O_5 oxide and a peak at 34.34° related to the ZnO (002) reflection are detected. In Fig. 4, only the most representative XRD profiles of the untreated CdTe and of the samples annealed with 40 mbar HCF_2Cl partial pressure were shown.
The preferential orientation of each film is analyzed by using the texture coefficient C_{hkl}, calculated by means of the following formula (Barret & Massalski 1980):

$$C_{hkl} = \frac{I_{hkl}/I_{hkl}^0}{\frac{1}{N}\sum_N I_{hkl}/I_{hkl}^0}, \tag{4}$$

where I_{hkl} is the detected intensity of a generic peak in the XRD spectra, I^0_{hkl} is the intensity of the corresponding peak for a completely randomly oriented CdTe powder (values taken from the JCPDS) and N the number of reflections considered in the calculation. C_{hkl} values above the unity represented a preferential orientation along the crystallographic direction indicated by the hkl indices. The texture coefficients C_{111}, calculated by the formula Nr 4 for all the samples, are summarized in table 2, together with the CL intensity ratios. A better comprehension of the orientation of each thin film as a whole can be obtained by the standard deviation σ of the C_{hkl} coefficients. Each value has been calculated by the following formula:

$$\sigma = \sqrt{\frac{\sum_N (C_{hkl} - 1)^2}{N}} \tag{5}$$

A complete randomly oriented film is expected to have a σ value as close as possible to 0.
The untreated CdTe thin film shows the highest preferential orientation along the (111) direction with a texture coefficient C_{111}=2.02. The effect of HCF$_2$Cl treatment is highlighted by a decrease of the (111) related intensity and by an increase of the relative intensities of the additional reflections (220), (311), (400), (331), (422) and (511), detected. The calculated σ value for the untreated CdTe is also the highest one (σ=0.52) demonstrating the oriented status of that film. This behavior is evidenced in Fig. 5, in which the calculated peak intensity ratios between each (220), (311), (400), (331), (422) and (511) additional reflection and the (111) one are plotted.
The combined effect of HCF$_2$Cl partial pressure and the total gas pressure, in the annealing chamber, could be also evidenced by comparing the C_{111} and σ values of the CdTe films treated by 30mbar HCF$_2$Cl, but higher total pressure (800mbar), sample F30H in table 2. Its values were higher than the CdTe treated with the same partial pressure and lower total pressure (sample F30L in table 1), but similar to the untreated CdTe.

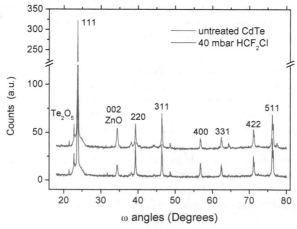

Fig. 4. XRD profiles of the untreated CdTe thin film compared to the sample annealed with 400 mbar Ar+Freon total pressure in the annealing chamber and 40mbar HCF$_2$Cl partial pressure.

Sample	XRD results		Morphology	1.4 eV/NBE CL intensity	
	C_{111} texture coefficient	σ	Average grain size (µm)	12 keV	25 keV
UT	2,02	0,52	11.7	0.9	0.72
F30L	1,12	0,29		2.3	3.39
F30H	1,7	0,42	10.8		10.75
F40	1,15	0,31	11.2	14.48	15.47
F50	0,56	0,36		14.74	19.97

Table 2. Summary of the results obtained by processing the XRD profiles, CL spectra and SEM images.

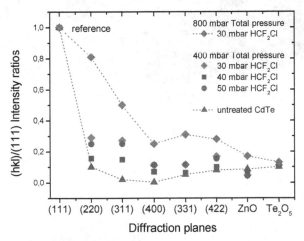

Fig. 5. Plot of intensity ratios among each diffraction (220), (311), (400), (331), (422) and (511) and the (111) one for all the studied CdTe thin films

The loss of preferential orientation due to HCF_2Cl annealing results in a slight modification of the CdTe morphology after the thermal treatment. The untreated CdTe films showed already large grains, as visible in the SEM image of Fig. 6 a. The average grain size obtained by processing the images was 11.7µm and the largest grains reached 20.4µm. The material treated with 40mbar HCF_2Cl partial pressure showed grains with dimensions similar (avg = 11.2µm) to those of the untreated one (Fig. 6 b). The observed average size confirmed that CSS grown CdTe did not show grain size increase after annealing in presence of chlorine as already described in the literature by several authors (Moutinho et al. 1998). Grain dimensions distribution extracted from the SEM images has been represented in histograms showed in Fig. 7 a and b. It could be observed that the small grains density in the HCF_2Cl treated material was reduced, producing a thinner distribution of the histogram columns.

On the contrary, all the Freon treated CdTe showed a remarkable grain shape variation with respect to the untreated sample where most of the grains appeared as tetragonal pyramids with the vertex aligned on the growth direction (Fig. 8 a). This shape justified their high preferential orientation along the (111) direction. This grain shape appeared clearly modified in the HCF_2Cl annealed films. They were more rounded and the pyramids seem to be made up by a superposition of "terraces" (Fig. 6 b). This morphology change could be correlated to the C_{111} texture coefficient decrease. Two possible mechanisms related to the HCF_2Cl annealing could be invoked: a re-crystallization effect or an "etching-like" erosion of the grain surface. The unmodified grain size and the appearance of the terraces seem to indicate that the latter phenomenon occurred during the Freon treatment.

Fig. 6. SEM image of the polycrystalline CdTe surface morphology: a) untreated film; b) annealed with 40mbar HCF_2Cl

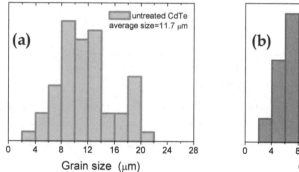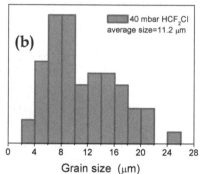

Fig. 7. Histograms of the grain size as obtained from the SEM images: a) untreated CdTe; b) CdTe annealed by 40mbar HCF_2Cl partial pressure.

The effect of thermal treatment on the CdTe bulk electro-optical properties has been studied by acquiring CL spectra at electron beam energy (E_B) of 25keV, corresponding to a maximum penetration depth of about 2.5μm. The CL generation volume dimensions were calculated by means of a numerical approach based on random walk Monte Carlo simulation developed in our laboratory (Grillo et al. 2003). The low temperature (77 K) spectrum of a 240x180 μm² region of the untreated CdTe showed the clear near bend edge (NBE) emission centered at 1.57eV. The temperature is too high to discriminate the acceptor from the donor bound excitonic line, we supposed they were superimposed underneath the NBE band. In addition to the NBE emission, two weak bands, centered at 1.47eV and 1.35eV respectively, were also detected. The 1.35eV and 1.47eV CL peaks were visible only in the untreated CdTe and their origin was not related to the HCF_2Cl treatment. The 1.35eV

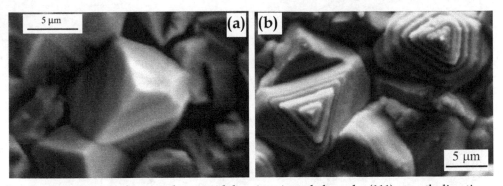

Fig. 8. a) SEM image of a typical pyramidal grain oriented along the (111) growth direction of the untreated film; b) SEM image showing pyramidal grains with terraces of the CdTe annealed by 40mbar HCF₂Cl partial pressure.

emission could be attributed to radiative recombination levels induced by impurities, like Cu, unintentionally incorporated during the CdTe deposition, or diffused from the front contact and buffer layers during the high temperature growth. The 1.47eV peak has been previously observed in polycrystalline CdTe (Cárdenas-García et al. 2005) and ascribed to the dislocation related Y-emission. In our untreated material, a clear dependence of this emission on the dislocations has not been demonstrated, but the disappearance of this peak in the annealed, high crystalline quality CdTe supports this attribution (Armani et al. 2007).

The HCF₂Cl annealing effect on the CdTe recombination mechanisms was studied by both CL spectroscopy and monochromatic (monoCL) mapping. CL spectra showed a drastic difference between untreated and HCF₂Cl annealed samples, as visible in Fig. 9. All the HCF₂Cl treated samples showed, in addition to the NBE emission, a broad CL band centered at 1.4eV which intensity increased by increasing the HCF₂Cl partial pressure, suggesting a strong dependence of this emission on the annealing. The literature studies on both single–crystal and polycrystalline CdTe (Consonni et al. 2006; Krustok et al. 1997) showed photoluminescence (PL) and CL bands centered at energies close to 1.4eV; their origin was attributed to a radiative recombination center like the well known A-center, due to a complex between a Cd vacancy (V_{Cd}) and a Cl impurity, in Cl-doped CdTe (Meyer et al. 1992; Stadler et al. 1995). The clear correlation between the 1.4eV band and the HCF₂Cl treatment supported the attribution of the 1.4eV band observed in our CdTe films to a complex like the A-centre. Either Cl or F impurities could be the origin of the level responsible for this transition. Several impurities, among which Cl and F, created acceptor levels with very similar energy values above the valence band edge as reported by Stadler et al. (Stadler W. et al. 1995). In particular the levels due to Cl and F differ solely by 9meV. The CL spectral resolution, lower than the PL one, did not allow determining the exact energy position of the 1.4 eV band with a precision better than 0.01eV. On this basis a clear attribution, to Cl or F, of the impurity creating the complex together to the V_{Cd} was impossible. The 1.4eV/NBE CL intensity ratios represented a tool to study the concentration of the V_{Cd}-Cl(F) complex responsible for the 1.4eV band; the comparison among the untreated and the annealed CdTe results obtained at 25keV have been summarized in Fig. 10.

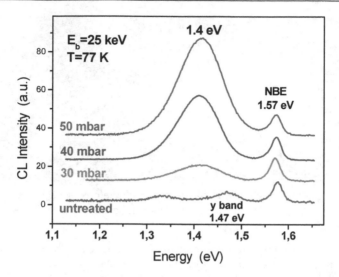

Fig. 9. Comparison among the low temperature (77 K) CL spectra (E_b=25keV) of untreated CdTe and samples annealed at a HCF_2Cl partial pressure of 30, 40 and 50 mbar.

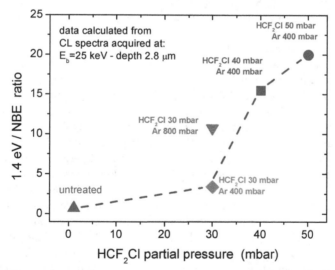

Fig. 10. Plot of the 1.4eV/NBE integrated intensity ratios. The experimental points have been calculated from spectra acquired in various regions of each CdTe film at E_b= 25keV.

One of the peculiarities of the CL technique is the possibility to increase the probing depth within the studied materials, by increasing E_B, by keeping the injection density in the generation volume constant. The effect of annealing on the CdTe luminescence behavior was expected to be more effective close to the CdTe surface. For this reason, CL spectra at lower beam energy (E_B=12keV), corresponding to a maximum penetration depth of 900nm below the CdTe surface, were collected. The comparison among untreated and annealed samples showed a dependence of the 1.4eV emission intensity on the HCF$_2$Cl partial pressure similar to the highly depth 25 keV analysis. A more detailed depth-resolved study of the V$_{Cd}$-Cl(F) complex distribution is performed by acquiring CL spectra at different E$_B$ from 8 to 36 keV, corresponding to a probing depth between 0,36 and 4,6 μm. Fig. 11 showed the CL spectra, normalized to the NBE intensity, in order to better highlight the 1.4eV band intensity variations. The 1.4eV emission intensity decreased till the generation volume of the CL signal extended to about 2μm, then kept almost constant for the following 2μm. This behavior could be more clearly appreciated in the inset of the figure where the 1.4eV/NBE integrated intensity ratio has been shown. Possible influence of material quality inhomogeneities along the deposition axis could be neglected because the NBE peak position did not change in the studied depth range as well as its intensity showed very small and random variations. The V$_{Cd}$-Cl(F) complex density was the highest close to the CdTe surface, due to the maximum effectiveness of the HCF$_2$Cl treatment however it decreased in depth, but not disappeared in the first 4μm of the film.

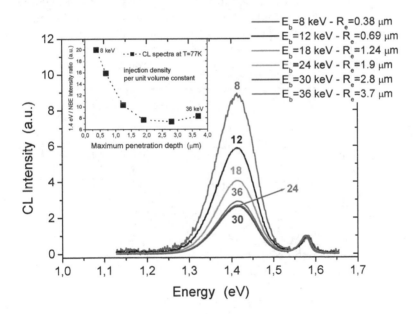

Fig. 11. Depth-dependent CL spectra acquired on the 40 mbar HCF$_2$Cl partial pressure annealed CdTe, by increasing E$_b$ from 8 to 36 keV; in the inset the plot of the 1.4 eV/NBE integrated intensity ratios as a function of E$_B$ has been shown.

The maximum E$_B$ suitable for the SEM used for this work (36keV) was not high enough to investigate the whole CdTe film, limiting the results to about a half of the material thickness. For this reason the samples were etched in order to eliminate a portion of CdTe, leaving the material near the CdTe/CdS interface free. After the etching procedure the decreasing thickness of CdTe film edge showed a slightly sloped surface extending from the upper surface of the front contact, down to about 1μm above the CdTe/CdS interface. The SEM image of the etched region has been shown in Fig. 13 a. CL analyses have been performed on the beveled CdTe surface, in the region closer to the interface, indicated as the "maximum etched surface" in Fig. 13 a. The comparison among the untreated CdTe and the annealed specimens was shown in Fig. 12. In the untreated and 30mbar treated samples, in addition to the NBE emission only the CL band centered at 1.47 eV has been observed. The 1.4eV band appeared only in cells treated with more than 30 mbar HCF_2Cl. By increasing the HCF_2Cl partial pressure, the 1.4eV CL intensity also increases confirming the behavior observed on the not-etched CdTe surface. On the other hand, when present, the 1.4eV CL intensities were lower than those observed on the not-etched surfaces and did not exceed the NBE intensity. This behavior was clear in Fig. 13 b, where the comparison among CL spectra acquired on the 40mbar HCF_2Cl etched surface at different depths approaching the CdTe/CdS interface was shown. The spectra were acquired from a small area (10µm wide) in order to investigate regions at the same depth. The analyzed regions have been indicated by the colored squares in Fig. 13 a. The 1.4eV CL intensity decreased remarkably from the blue to the black curve, which corresponded to regions approaching the "maximum etched region". Only in the spectrum acquired 3µm far from this region the 1.4eV band was back the dominant one. The depth-dependent decrease of the 1.4eV CL intensity could be attributed to a not-uniform distribution of the V_{Cd}-Cl(F) complex responsible for that transition because of a low diffusion of the Cl (or F) atoms within the CdTe.

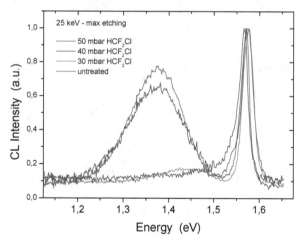

Fig. 12. Comparison among the low temperature (77 K) CL spectra (E_B=25 keV) of untreated CdTe and samples annealed at a HCF_2Cl partial pressure of 30, 40 and 50 mbar.

The spatial distribution of luminescence properties of CdTe was studied by acquiring monoCL images at the emission energies of the bands observed in the CL spectra, 1.57eV and 1.4eV respectively. The monoCL image collected at the NBE emission energy (E=1.57eV) showed the maximum intensity contribution from the central part of the grains (Fig. 14 b). The excitonic transitions came mainly from the CdTe grains, meaning that they were of good crystalline quality. On the other hand, in the same image the boundary regions between adjacent grains (*grain boundaries*) showed a dark contrast which corresponded to a very low radiative recombination efficiency. By acquiring the monoCL image at E=1.4eV

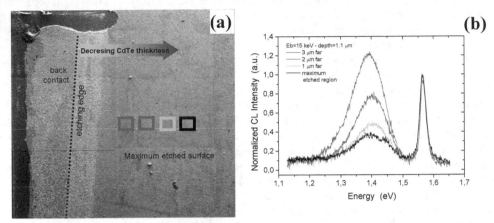

Fig. 13. a) SEM image of the etched surface of the solar cell annealed with 40 mbar HCF$_2$Cl partial pressure; b) comparison among the low temperature (77K) CL spectra (E$_b$=15keV) of the 40 mbar HCF$_2$Cl partial pressure etched solar cell.

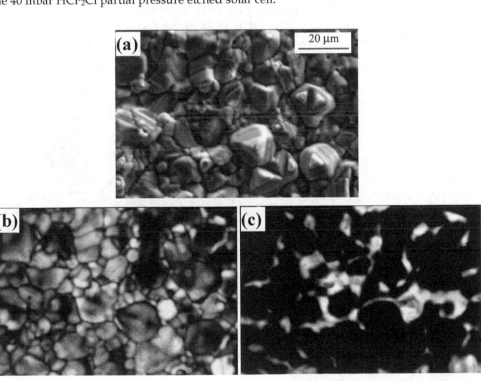

Fig. 14. 40 mbar HCF$_2$Cl partial pressure CdTe CL mapping; a) SEM image of the surface morphology; b) monoCL image at the NBE emission energy (E=1.57eV); c) monoCL at E=1.4eV emission energy.

(Fig.14 c) a complementary CL intensity distribution has been observed. The bright contrast was concentrated in the grain boundaries. For a more clear representation of the correlation between surface morphology and CdTe radiative recombination properties the corresponding SEM image has been shown (Fig. 14 a).

Taking advantage on the possibility of focusing the SEM electron beam in suitable small regions, the spectroscopic behavior of the CL in a single grain has been studied. The adopted investigation conditions were in this case "fixed beam" at $E_B=12keV$, which corresponded to a sub-micrometric generation volume. The CL spectra acquired in the points marked by numbered colored circles on Fig. 15 a have been shown in Fig. 15 b. The curves were normalized to the NBE emission intensity. The spectra collected inside the CdTe grain numbered 2-5 showed a high-intensity NBE emission dominating the spectra. The 1.4eV band intensity, on the contrary, exceeded the NBE one in the spectra 1 and 6, collected close to the grain edges (*grain boundary*). The spatial distribution of the 1.4eV emission intensity could be directly correlated to the non homogenous density of the V_{Cd}-Cl(F) complex responsible for that emission. A gettering of the native point defects (V_{Cd}), as well as of the incorporated impurities (Cl or F) could be suggested as the origin of this inhomogeneity. A preferential diffusion through the grain boundaries, of those impurities, during the high temperature annealing, has been supposed. The correlation between polycrystalline Cl doped CdTe structural properties and electronic levels created by the dopants in the CdTe gap has been discussed in the literature (Consonni et al. 2006, 2007) and the published results were in good agreement to those presented in this work.

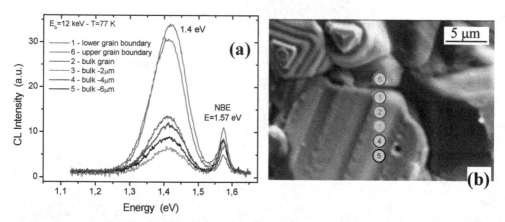

Fig. 15. a) Spatial variation of CL emissions, by acquiring the spectra in spot mode, in different points of a single CdTe grain; b) SEM image of the region where the CL spots are collected.

4.2 Comparison between the material properties and device performances

The behavior of the electro-optical properties of CdTe as a function of HCF_2Cl partial pressure, shown in the previous paragraph, did not allow to obtain information about the role of HCF_2Cl on the cell photovoltaics parameters and to investigate the transport processes taking place at the junction. The electrical properties of the annealed solar cells were compared to those of an untreated device. Dark reverse I-V curves were shown in Fig. 16 a, where the evolution of the reverse current (I_0) as a function of the HCF_2Cl partial pressure can be observed.

The diode ideality factor (A) has been calculated from those curves and its behavior as function of HCF_2Cl was also reported in Fig. 16 b. Specific processes occurring at the junction determined the reverse current and diode factor. In our case, it was observed a decrease of the reverse current when the HCF_2Cl partial pressure was increased. This behavior reached a minimum in the most efficient device obtained for this series, corresponding to 40mbar HCF_2Cl partial pressure (J_{sc}=26.2mA/cm^2, V_{oc}=820mV, ff=0.69, η=14.8%, see Fig. 17). An increase of 10mbar more reactive gas in the annealing chamber yields to a degradation of the reverse current that was increased of various orders of magnitude, showing the high reactivity of the treatment and the impact of an excess annealing on the device electrical performance. At the same time, from the behavior of A, a variation of transport mechanism depending on the treatment conditions could be suggested (Fig. 16 b). For the untreated sample, A=1.8 indicated that recombination current dominated the junction transport mechanism or that high injection conditions were present. An increase of the HCF_2Cl partial pressure gave rise to a situation in which diffusion and recombination currents take place together until the case of 40mbar HCF_2Cl partial pressure was reached, where the minimum value of A=1.2, appointed to a predominant diffusion current. The cell treated with 50mbar of reactive gas partial pressure showed a sharp modification, by increasing again the diode factor n up to 1.8. The increase of the diode reverse saturation current was responsible for a drastic reduction of ff (Fig16 b), despite the J_{SC} and V_{OC} did not change appreciably from the others HCF_2Cl annealed devices.

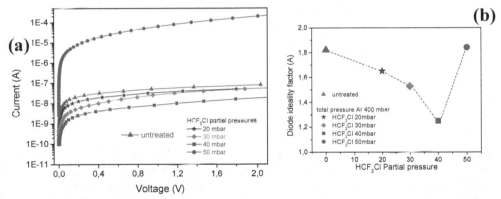

Fig. 16. a) Comparison among the dark reverse I-V curves for untreated and, 20, 30, 40 and 50 mbar of HCF_2Cl partial pressure treated solar cells; b) Diode ideality factor A as a function of the HCF_2Cl partial pressure.

The evolution of the J-V light curves (Fig. 17) of all samples showed an increase of the photovoltaic parameters by increasing the Freon partial pressure until 40mbar, while the J-V characteristic of the sample F50 showed a decrease of the fill factor to 0.25. The latter behavior could be related to a very strong intermixing between CdS and CdTe, due to the treatment, so that a very large p-n junction region was present.
A clear roll-over behavior of all the J-V curves was observed in the Fig. 17; mainly for the untreated sample and F20 and F50. This behavior was attributed to an n-p parasitic junction, opposite to the main p-n junction created by the back contact. We assume that this behavior was also strongly related to the incorporation of Cl impurities into CdTe. In our belief, the increment of the photocurrent collection should be essentially due to an increment of the

photogenerated minority carriers lifetime in the CdTe layer which suggested that the passivation of defects in absence of Cl contributed as non radiative recombination centers (Consonni et al. 2006). We considered the 50mbar HCF_2Cl cell an overtreated sample where the intermixing process was so strong that all the available CdS was consumed. The presence of shunt paths through the junction can explain the high reverse current and low fill factor values.

The luminescence properties observed on the CdTe material showed a continuous increase of the 1.4eV band intensity as a function of HCF_2Cl partial pressure; the device electrical characterization showed, on the contrary, a threshold at 40mbar partial pressure. Above this value the solar cell performances collapsed dramatically suggesting a critical correlation between HCF_2Cl annealing and junction properties.

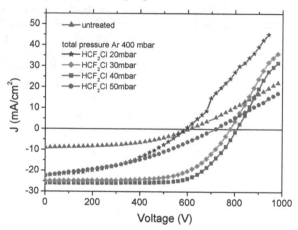

Fig. 17. Room temperature I-V characteristics under AM 1.5, 100mW/cm^2 illumination conditions of untreated solar cells compared to the 20, 30, 40 and 50 mbar HCF_2Cl partial pressures respectively.

The comparison between the diode factor A and the 1.4eV intensity behaviors suggested that the V_{Cd}-Cl(F) complex was beneficial for the device performances, but did not explain alone the maximum efficiency value measured for the 40 mbar annealed solar cells. A combined CdTe material doping and grain boundaries passivation effect had to be invoked. The absence of the 1.4eV band in the untreated and low HCF_2Cl partial pressure annealed CdTe after etching demonstrated that a non-radiative recombination centre was responsible for the low A values. This centre was then passivated by the Cl (or F) incorporation till the excess, for HCF_2Cl partial pressures above 40 mbar, deteriorated the p-n junction.

The complex V_{Cd}-Cl(F) formation could also be supported by the temperature dependent I-V analyses carried out on the CdTe thin film. The Arrhenius plot extracted from the CdTe dark conductivity as a function of the inverse of the temperature has been shown in Fig.18. The plot showed that, in the case of untreated CdTe the high calculated activation energy (324meV) has been related to a level due to the presence of occasional impurities like Cu, Ag or Au; the activation energy decreases by increasing the HCF_2Cl partial pressure, down to E_a=142meV for the material treated by 40mbar HCF_2Cl partial pressure. This value was in good agreement with those obtained in Cl (or F) doped CdTe single-crystals and attributed to the A-centre, due to the complex V_{Cd}-Cl(F) acceptor-like (Meyer et al. 1992).

A model of the effect of annealing as a function of HCF$_2$Cl partial pressure, on the bulk CdTe and its grain boundaries as well as on the CdTe-CdS intermixing mechanisms occurring at the interface has been showed in Fig. 19. The Cl (or F) impurities contained in the annealing gas penetrate into the material partially doping the CdTe. The major part was gettered to the grain boundaries, as observed in the monoCL image (Fig. 14 c), passivating them and improving conductivity. Contemporary the interdiffusion of S in the CdTe and of Te in CdS has been promoted by creating an intermixing region, which thickness increased by increasing the HCF$_2$Cl partial pressure, pictured by the orange region between CdTe and CdS. The poor solar cell performances of the 50mbar HCF$_2$Cl partial pressure annealed device have been explained by a complete consumption of the CdS layer and by destruction of the main p-n junction.

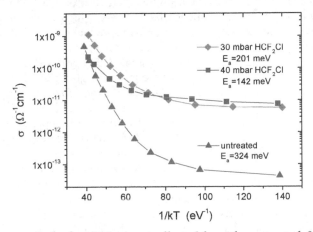

Fig. 18. Temperature dependent I-V curves collected from the untreated, 30mbar and 40mbar HCF$_2$Cl partial pressures respectively.

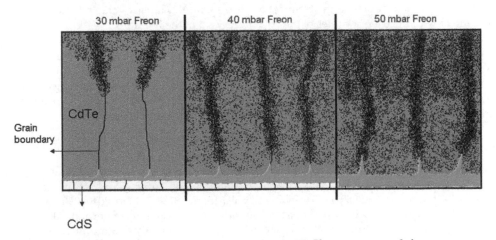

Fig. 19. Schematic representation of the effect of the HCF$_2$Cl treatment on defects distribution and intermixing junction formation

5. Conclusions

Thin films CdTe deposited by CSS have been submitted to a novel, full dry, post-deposition treatment based on HCF_2Cl gas. The annealing demonstrated to affect the structural properties of the materials through the loss of preferential orientation. Texture coefficient of the (111) Bragg reflection decreased from 2, for the untreated CdTe, down to 0.56 for the film treated with the highest HCF_2Cl partial pressure. On the contrary, the grain size did not show any change after annealing maintaining an average dimension of about 12μm. These results were common for high temperature CSS deposited CdTe films, while a clear dependence on the HCF_2Cl partial pressure of the electro-optical properties of the films have been observed through the presence of a 1.4 eV CL band in the annealed specimens. The transition responsible for this emission involved an electronic level in the gap with an energy of about 0.15 eV above the valence band edge, which could be attributed to a complex between cadmium vacancy and an impurity probably identified in Cl or F (V_{Cd}-Cl/F) from the annealing gas.

The combined CL mapping and spectroscopy on single CdTe grains showed that the lateral distribution of this complex was not homogeneous in the grain, but it was concentrated close to the grain boundaries. The bulk grain, on the contrary, showed a high optical quality, evidenced by the predominance of the NBE emission. The in-depth effectiveness of the HCF_2Cl annealing has been demonstrated by correlating depth-dependent CL analyses to the study of the beveled CdTe surface due to the Br-methanol etching. High density of the V_{Cd}-Cl/F complex responsible for the 1.4 eV band has been observed close to the CdTe surface; it decreased by increasing depth in the bulk region of the film about 5μm below the surface. By removing several microns of CdTe material and by approaching the CdTe/CdS interface, in the etched specimens, an HCF_2Cl partial pressure higher than 30 mbar was necessary to detect the 1.4 eV emission, this means to create the V_{Cd}-Cl/F complex. On the other hand electrical characterization determined a threshold in the beneficial role of the HCF_2Cl annealing, showing the best solar cell performances for the 40 mbar partial pressure treated device. Temperature dependent I-V analyses showed a remarkable decrease of the electronic level activation energy, from 348meV to 142meV. The last value resulted in good agreement with the energy values of the A-center found in the literature.

The comparison between the diode factor A and the 1.4 eV CL band intensity behaviors evidenced that the V_{Cd}-Cl/F complex was beneficial for the device performance, but does not explain alone the maximum efficiency value measured for the 40 mbar annealed solar cells. A tentative schematic model of the mechanisms occurring during post-deposition treatment, in the bulk CdTe and close to the CdTe/CdS interface have been also proposed. A combined CdTe-CdS intermixing and grain boundaries passivation effect has to be invoked.

6. References

Armani N., Salviati G., Nasi L., Bosio A., Mazzamuto S. and Romeo N., "Role of thermal treatment on the luminescence properties of CdTe thin films for photovoltaic applications", (2007) *Thin Solid Films*, vol. 515, pp. 6184-7, ISSN 00406090

Barret C. and Massalski T.B. Structure of Metals, edited by Pergamon, Oxford, p. 204 (1980)

Barrioz, V.; Lamb, D.A.; Jones; E.W. & Irvine, S.J.C. (2010). Suitability of atmospheric-pressure MOCVD CdTe solar cells for inline production scale. *Materials Research Society Symposium Proceedings*; ISBN: 978-160511138-4; San Francisco, CA; April 2009

Bätzner, D.L.; Romeo, A.; Zogg, H.; Wendt, R. & Tiwari, A.N. (2001) Development of efficient and stable back contacts on CdTe/CdS solar cells. *Thin Solid Films*, Vol.387, No.1-2, May 2001, pp. 151-154. ISSN: 00406090.

Birkmire, R.W.; Meyers, P.V. (1994) Processing issues for thin-film CdTe cells and modules, *Proceedings of the 24th IEEE Photovoltaic Specialists Conference*, ISSN: 01608371, Waikoloa, HI, USA, December 1994.

Bosio, A.; Romeo, N.; Mazzamuto, S. and Canevari V. (2006), Polycrystalline CdTe thin films for photovoltaic applications. *Progress in Crystal Growth and Characterization of Materials*, Vol.52, No.4, December 2006, pp. 247-279, ISSN: 09608974.

Aguilar-Hernandez J, Sastre-Hernandez J, Mendoza-Perez R, Cardenas-Garcia M and Contreras-Puente G (2005), "Influence of the CdCl2 thermal annealing on the luminescent properties of CdS-CSVT thin films", *Physica Status Solidi C - Conference and Critical Reviews*, vol. 2, No 10, p.p. 3710-3713, ISSN: 1610-1634

Chu, T.L.; Chu, S.S.; Ferekides, C.; Wu, C.Q.; Britt, J. & Wang, C. (1991). 13.4% efficient thin-film CdS/CdTe solar cells. *Journal of Applied Physics*, Vol.70, No.12, 1991, pp. 7608-7612. ISSN: 00218979

Compaan, A.D.; Tabory, C.N.; Li, Y.; Feng, Z. & Fischer, A. (1993). CdS/CdTe Solar cells by RF sputtering and by laser physical vapor deposition. *Proceedings of the 23rd IEEE Photovoltaic Specialists Conference*, pp. 394-399, ISBN: 0780312201, Louisville, KY, USA; May 10-14, 1993

Consonni V., Feuillet G. and Renet S., (2006), "Spectroscopic analysis of defects in chlorine doped polycrystalline CdTe" *J. Appl. Phys.*, vol. 99, p.p. 053502-1/7, ISSN: 0021-8979

Consonni V., Feuillet G., Bleuse j. and Donatini F., (2007) "Effects of island coalescence on the compensation mechanisms in chlorine doped polycrystalline CdTe", *J. Appl. Phys.*, vol. 101, p.p- 063522-1/6, ISSN: 0021-8979

Cruz, L.R.; Kazmerski, L.L.; Moutinho, H.R.; Hasoon, F.; Dhere, R.G. & De Avillez, R. (1999). Influence of post-deposition treatment on the physical properties of CdTe films deposited by stacked elemental layer processing. *Thin Solid Films*, Vol.350, No. 1, August 1999, pp. 44-48. ISSN: 00406090

Cunningham, D.; Davies, K.; Grammond, L.; Mopas, E.; O'Connor, M.; Rubcich, M; Sadeghi, M.; Skinner, D. and Trumbly, T. (2000). Large area Apollo module performance and reliability. *Proceedings of the 28th IEEE Photovoltaic Specialists Conference*.

Enriquez, J.P. and Mathew, X. (2004). XRD study of the grain growth in CdTe films annealed at different temperatures. *Solar Energy Materials and Solar Cells*. Vol. 81, No. 3, February 2004, pp. 363-369, ISSN: 09270248

Ferekides, C.S.; Marinskiy, D.; Viswanathan, V.; Tetali, B.; Palekis, V.; Selvaraj, P. & Morel, D.L. (2000). High efficiency CSS CdTe solar cells. *Thin Solid Films*. Vol.361, (February 2000), pp. 520-526, ISSN: 00406090

Grillo V., Armani N., Rossi F., Salviati G. and Yamamoto N., (2003), "Al.L.E.S.: A random walk simulation approach to cathodoluminescence processes in semiconductors" Inst. Phys. Conf. Ser. No 180, p.p. 559-562, ISSN 0951-3248

Hartley, A.; Irvine, S.J.C.; Halliday, D.P. & Potter, M.D.G. (2001). The influence of CdTe growth ambient on MOCVD grown CdS/CdTe photovoltaic cells. *Thin Solid Films*. Vol.387, No.1-2, May 2001, pp. 89-91. ISSN: 00406090

Hernández-Contreras, H.; Contreras-Puente, G. , Aguilar-Hernández, J. , Morales-Acevedo, A. , Vidal-Larramendi, J. & Vigil-Galán, O. (2002). CdS and CdTe large area thin films processed by radio-frequency planar-magnetron sputtering. *Thin Solid Films*. Vol.403-404, February 2002, Pages 148-152. ISSN: 00406090

Josell, D.; Beauchamp, C. R.; Jung, S.; Hamadani. B. H.; Motayed, A.; Richter, L. J.; Williams, M.; Bonevich, J. E.; Shapiro, A. ; Zhitenev, N. & Moffat, T. P. (2009). Three Dimensionally Structured CdTe Thin-Film Photovoltaic Devices with Self-Aligned Back-Contacts: Electrodeposition on Interdigitated Electrodes. *Journal of Electrochemical Society*. Vol.156, No.8, (June 2009). pp. H654-H660, ISSN: 0013-4651

Kosyachenko, L.A.; Mathew, X.; Motushchuk, V.V. & Sklyarchuk, V.M. (2006). Electrical properties of electrodeposited CdTe photovoltaic devices on metallic substrates: study using small area Au–CdTe contacts. *Solar Energy*. Vol.80, No.2, (February 2006), pp. 148-155, ISSN: 0038092X

Krustok, J; Madasson, J; Hjelt, K, and Collan H (1997) "1.4 eV photoluminescence in chlorine-doped polycrystalline CdTe with a high density of defects" *Journal of Materials Science*, vol. 32 No: 6 p.p. 1545-1550, ISSN 00222461

Lane, D.W.; Rogers, K.D.; Painter, J.D.; Wood, D.A. & Ozsan, M.E. (2000). Structural dynamics in CdS-CdTe thin films. *Thin Solid Films*, Vol.361, February 2000, pp. 1-8. ISSN: 00406090

Levy-Clement, C. (2008). Thin film electrochemical deposition at temperatures up to 180 °C for photovoltaic applications. *Proceedings of Interfacial Electrochemistry and Chemistry in High Temperature Media - 212th ECS Meeting*; ISBN: 978-160560312-4; Washington, DC; October 2007

Lincot, D. (2005). Electrodeposition of chalcogenide semiconductors. *Proceedings of 207th Electrochemical Society Meeting*, Quebec, May 2005.

Luschitz, J.; Siepchen, B.; Schaffner, J.; Lakus-Wollny, K.; Haindl, G.; Klein, A. and Jaegermann, W. (2009) CdTe thin film solar cells: Interrelation of nucleation, structure, and performance. *Thin Solid Films*, Vol. 517, No.7, February 2009, pp. 2125-2131, ISSN: 00406090.

McCandless, B.E. & Birkmire, R.W. (1991) Analysis of post deposition processing for CdTe/CdS thin film solar cells. *Solar Cells*, vol. 31, No. 6, December 1991, pp. 527-535, ISSN: 03796787.

McCandless, B.E.; Moulton, L.V. and Birkmire, R.W. (1997). Recrystallization and sulfur diffusion in CdCl₂-treated CdTe/CdS thin films. *Progress in Photovoltaics: Research and Applications*, Vol. 5, No.4, July 1997, pp. 249-260, ISSN: 10627995.

McCandless, B.E. (2001). Thermochemical and kinetic aspects of cadmium telluride solar cell processing. *MRS Proceedings*, Vol.668, H1.6, ISSN: 02729172, San Francisco, CA, USA; April 16-20, 2001

McCandless, B.E. & Sites, J.R. (2003) Cadmium telluride solar cells in: *Handbook of Photovoltaic Science and Engineering* Luque, A. and Hegedus, S.S., pp. 617–662, John Wiley & Sons Ltd, ISBN: 9780471491965, Chichester, UK.

McCandless, B.E. and Dobson, K.D. (2004). Processing options for CdTe thin film solar cells. *Solar Energy*, Vol. 77, No.6, December 2004, pp. 839-856, ISSN: 0038092X.

Merano, M.; Sonderegger, S.; Crottini, A.; Collin, S.; Pelucchi, E.; Renucci, P.; Malko, A.; Baier, M.H.; Kapon, E.; Ganiere, J.D. and Deveaud, B. (2006) Time-resolved cathodoluminescence of InGaAs/AlGaAs tetrahedral pyramidal quantum structures. *Applied Physics B: Lasers and Optics*, Vol. 84, No.1-2, July 2006, pp. 343-350, ISSN: 09462171.

Meyer BK, Stadler W, Hofmann DM, Ömling P, Sinerius D and Benz KW (1992), On the Nature of the Deep 1.4 eV Emission Bands in CdTe - a Study with Photoluminescence and ODMR Spectroscopy, *Journal of Crystal Growth*, vol. 117, No 1-4, p.p. 656-659 ISSN: 0022-0248

Moutinho, H.R.; Al-Jassim, M.M.; Levi, D.H.; Dippo, P.C. and Kazmerski, L.L. (1998). Effects of CdCl$_2$ treatment on the recrystallization and electro-optical properties of CdTe thin films. *Journal of Vacuum Science and Technology A: Vacuum, Surfaces and Films*. Vol.16, No.3, May 1998, pp. 1251-1257, ISSN: 07342101.

Moutinho, H.R.; Dhere, R.G.; Al-Jassim, M.M.; Levi D.H. and Kazmerski L.L. (1999). Investigation of induced recrystallization and stress in close-spaced sublimated and radio-frequency magnetron sputtered CdTe thin films. *Journal of Vacuum Science and Technology A: Vacuum, Surfaces and Films*. Vol.17, No.4, 1999, pp. 1793-1798, ISSN: 07342101.

Paulson P.D. and Dutta V (2000). Study of in situ CdCl2 treatment on CSS deposited CdTe films and CdS/CdTe solar cells, *Thin Solid Films*,vol. 370, p.p. 299-306, ISSN 00406090

Perrenaud, J, Kranz, L.; Buecheler S.; Pianezzi, F. and Tiwari, A.N. (2011). The use of aluminium doped ZnO as trasparent conductive oxide for CdS/CdTe solar cells. *Thin Solid Films*. Aricle in Press. ISSN: 00406090

Plotnikov, V., Liu, X., Paudel, N., Kwon, D., Wieland & K.A., Compaan, A.D. (2011). Thin-film CdTe cells: Reducing the CdTe. *Thin Solid Films*. Article in Press. ISSN: 00406090

Potlog, T.; Khrypunov, G.; Kaelin, M.; Zogg, H. & Tiwari, A.N. (2007). Characterization of CdS/CdTe solar cells fabricated by different processes. *Materials Research Society Symposium Proceedings*, ISBN: 978-155899972-5, San Francisco, CA; April 2007

Potter, M.D.G.; Halliday, D.P.; Cousins, M. & Durose, K. (2000). Study of the effects of varying cadmium chloride treatment on the luminescent properties of CdTe/CdS thin film solar cells. *Thin Solid Films*, Vol.361, February 2000, pp. 248-252. ISSN: 00406090

Romeo, A.; Bätzner, D.L.; Zogg, H. & Tiwari, A.N. (2000). Recrystallization in CdTe/CdS. *Thin Solid Films*. Vol.361 (February 2000), pp. 420-425, ISSN: 00406090

Romeo, A.; Buecheler, S.; Giarola, M.; Mariotto, G.; Tiwari, A.N.; Romeo, N.; Bosio, A. & Mazzamuto, S. (2009). Study of CSS- and HVE-CdTe by different recrystallization processes. *Thin Solid Films*, Vol.517, No.7, February 2009, pp. 2132-2135. ISSN: 00406090

Romeo, N.; Bosio, A.; Tedeschi, R.; Romeo, A. & Canevari, V. (1999). Highly efficient and stable CdTe/CdS thin film solar cell. *Solar Energy Materials and Solar Cells*, Vol.58, No.2, June 1999, pp.209-218. ISSN: 09270248

Romeo, N.; Bosio, A.; Canevari, V. & Podestà, A. (2004). Recent progress on CdTe/CdS thin film solar cells. *Solar Energy*, Vol.77, No.6, December 2004, pp. 795-801. ISSN: 0038092X

Romeo, N.; Bosio, A.; Mazzamuto, S.; Podestà, A. and Canevari, V. (2005). The Role of Single Layers in the Performance of CdTe/CdS thin film solar cells. *Proceedings of 25th Photovoltaic Energy Conference and Exhibition*,Barcelona, Spain, June 2005.

Romeo, N.; Bosio, A.; Romeo, A.; Mazzamuto, S. & Canevari, V. (2006). High Efficiency CdTe/CdS Thin Film Solar Cells Prepared by Treating CdTe Films with a Freon Gas in Substitution of CdCl$_2$. *Proceedings of the 21st European Photovoltaic Solar Energy Conference and Exhibition*, pp.1857-1860, ISBN 3-936338-20-5, Dresden, Germany, September 4-8, 2006.

Romeo, N.; Bosio, A.; Mazzamuto, S.; Romeo, A. & Vaillant-Roca, L. (2007). "High efficiency cdte/cds thin film solar cells with a novel back-contact", *Proceedings of the 22nd European Photovoltaic Solar Energy Conference and Exhibition*, pp. 1919-1921, Milan; Italy.

Romeo, N.; Bosio, A.; Romeo, A. & Mazzamuto, S. (2010). A CdTe thin film module factory with a novel process. *Proceedings of 2009 MRS Spring Meeting;* Vol.1165, pp. 263-273, ISBN: 978-160511138-4, San Francisco, CA, USA; April 13-17, 2009

Schulz, D.L.; Pehnt, M.; Rose, D.H.; Urgiles, E.; Cahill, A.F.; Niles, D.W.; Jones, K.M.; Ellingson, R.J.; Curtis, C.J.; Ginley, D.S. (1997). CdTe Thin Films from Nanoparticle Precursors by Spray Deposition. *Chemistry of Materials.* Vol.9. No.4, April 1997, pp. 889-900. ISSN: 08974756

Stadler, W.; Hoffmann, D.M.; Alt, H.C.; Muschik, T.; Meyer, B.K.; Weigel, E.; Müller-Vogt, G.; Salk, M.; Rupp, E. and Benz, K.W. (1995) Optical investigations of defects in $Cd_{1-x}Zn_xTe$, *Physical Review B*, vol.51, No.16 1995, pp. 10619-10630, ISSN: 01631829.

Sze, S. (1981). *Physics of Semiconductor Devices* (2nd ed.), Wiley, ISBN:9780471143239, New York.

Yacobi, B.G. & Holt, D.B. (1990) *Cathodoluminescence microscopy of inorganic solids*, Plenum Press, ISBN: 0306433141, New York and London.

Yi, X. & Liou, J.J. (1995). Surface oxidation of polycrystalline cadmium telluride thin films for Schottky barrier junction solar cells. *Solid-State Electronics.* Vol.38, No.6, (1995), pp.1151-1154, ISSN: 00381101.

Yoshida, T. (1992). Analysis of photocurrent in screen-printed CdS/CdTe solar cells. *Journal of the Electrochemical Society.* Vol.139, No.8, August 1992, pp. 2353-2357. ISSN: 00134651

Yoshida, T. (1995). Photovoltaic properties of screen-printed CdTe/CdS solar cells on indium-tin-oxide coated glass substrates. *Journal of the Electrochemical Society.* Vol.142, No.9, September 1995, pp. 3232-3237. ISSN: 00134651

Wu, X.; Keane, J.C.; Dhere, R.G.; DeHart, C.; Duda, A.; Gessert, T.A.; Asher, S.; Levi, D.H. and Sheldon, P. (2001 a). 16.5% efficient CdS/CdTe polycrystalline thin-film solar cell, *Proceedings of the 17th E-PVSEC*, München, Germany; October 2001.

Wu, X.; Asher, S.; Levi, D.H.; King, D.E.; Yan, Y.; Gessert, T.A. & Sheldon, P. (2001 b). Interdiffusion of CdS and Zn_2SnO_4 layers and its application in CdS/CdTe polycrystalline thin-film solar cells. *Journal of Applied Physics*, Vol.89, No.8, April 2001, pp. 4564-4569. ISSN: 00218979

Wu, X. (2004). High-efficiency polycrystalline CdTe thin-film solar cells. *Solar Energy.* Vol.77, No6, (December 2004), pp. 803-814, ISSN: 0038092X

Wu, X.; Zhou, J.; Duda, A.; Yan, Y.; Teeter, G.; Asher, S.; Metzger, W.K.; Demtsu, S.; Wei, S.-Huai. & Noufi, R. (2007). Phase control of Cu_xTe film and its effects on CdS/CdTe solar cell. *Thin Solid Films*, Vol.515, No.15 SPEC. ISS., May 2007, pp. 5798-5803, ISSN: 00406090

Xiaonan Li,. Niles D. W,. Hasoon F. S,. Matson R. J, and Sheldon P. (1999). Effect of nitric-phosphoric acid etches on material properties and back-contact formation of CdTe-based solar cells. *J. Vac. Sci. Technol. A*, vol. 17, No 3, p.p. 805-809 ISSN: 07342101.

Zanio, K.; Willardson R.K. & Beer, A.C. (1978). *Cadmium telluride. Volume 13 of Semiconductors and semimetals. Cadmium telluride*, Academic Press, ISBN 0127521135, 9780127521138, London, UK

Zhou, J.; Wu, X.; Duda, A.; Teeter, G. & Demtsu, S.H. (2007). The formation of different phases of CuxTe and their effects on CdTe/CdS solar cells. *Thin Solid Films*, Vol. 515, No.18, June 2007, pp. 7364-7369, ISSN: 00406090

Zoppi, G.; Durose, K.; Irvine, S.J.C. & Barrioz, V. (2006). Grain and crystal texture properties of absorber layers in MOCVD-grown CdTe/CdS solar cells. *Semiconductor Science and Technology.* Vol.21, No.6, June 2006, pp. 763-770. ISSN: 02681242

3

Computer Modeling of Heterojunction with Intrinsic Thin Layer "HIT" Solar Cells: Sensitivity Issues and Insights Gained

Antara Datta and Parsathi Chatterjee
Energy Research Unit, Indian Association for the Cultivation of Science,
Jadavpur, Kolkata,
India

1. Introduction

Despite significant progress in research, the energy provided by photovoltaic cells is still a small fraction of the world energy needs. This fraction could be considerably increased by lowering solar cell costs. To achieve this aim, we need to economize on the material and thermal budgets, as well as increase cell efficiency. The silicon "Heterojunction with Iintrinsic Thin layer (HIT)" solar cell is one of the promising options for a cost effective, high efficiency photovoltaic system. This is because in "HIT" cells the P/N junction and the back surface field (BSF) layer formation steps take place at a relatively low temperature (~200°C) using hydrogenated amorphous silicon (a-Si:H) deposition technology, whereas in normal crystalline silicon (c-Si) cells the wafer has to be raised to ~800°C for junction and BSF layer formation by diffusion. This means not only a lower thermal budget, but also cost reduction from thinner wafers, since the danger of the latter becoming brittle is strongly reduced at lower (~200°C) temperatures. Thin intrinsic layers on either face of the c-Si substrate, effectively passivate c-Si surface defects, which would otherwise degrade cell performance. Moreover it has been demonstrated that carriers can pass through the passivating layers without significant loss.

In this chapter, we use detailed electrical-optical modeling to understand carrier transport in these structures and the sensitivity of the solar cell output to various material and device parameters. The global electrical - optical model "Amorphous Semiconductor Device Modeling Program (ASDMP)", originally conceived to simulate the characteristics of solar cells based on disordered thin films, and later extended to model also mono-crystalline silicon and "HIT" solar cells (Nath et al, 2008), has been used for all simulations in this chapter. The model takes account of specular interference effects, when polished c-Si wafers are used, as well as of light-trapping when HIT cells are depositd on textured c-Si.

2. Historical development of HIT solar cells

One of the successful applications of hydrogenated amorphous silicon (a-Si:H) is in crystalline silicon heterojunction (HJ) solar cells. Fuhs et al (1974) first fabricated heterojunction silicon solar cells, where the absorber is P (N) type c-Si, while the emitter N

(P) a-Si:H layer is deposited by the standard plasma-enhanced chemical vapor deposition (PECVD) technique at ~200°C. However the efficiency achieved was much lower than in c-Si solar cells. In the early 80's Prof. Y. Hamakawa and his co-workers [Osuda et al, 1983] predicted the relevance of a-Si:H /c-Si stacked solar cells in silicon applications. Following the study of Prof. Hamakawa, many research groups world wide became interested in the technological development of a-Si:H/c-Si heterojuction solar cells as an alternative to traditional diffused emitter solar cells. It was almost a decade later that Sanyo began work in 1990 on the growth of low temperature junctions on c-Si and developed a new type of heterojunction solar cells called ACJ-HIT (Artificially Constructed Junction- Heterojunction with Intrinsic Thin layer), now shortened to "HIT", with a conversion efficiency of 18.1% (Tanaka et al, 1992) that has thereafter been continuously improved to yield an outstanding 22% efficiency in 100 cm² solar cells (Taguchi et al, 2005). Moreover Sanyo also achieved 19.5% efficiency in mass production (Tanaka et al, 2003). The innovation that made this possible was the introduction of thin films of intrinsic a-Si:H on either side of the c-Si wafer, to passivate the defects on its surface, that were responsible for the low efficiency of the earlier heterojunction cells [Fuhs et al, 1974]. A low recombination surface velocity of 15 cm/s has been demonstrated for passivation by intrinsic a-Si:H by Wang et al (2005). This is as good as the best dielectric surface passivation, such as by SiO_2 and amorphous silicon nitride (SiN_x) (Meier et al, 2007). More importantly, the a-Si:H I-layer can be inserted between the c-Si and a doped layer without significant restriction to carrier transport. The device structure of HIT cells that has been developed by Sanyo is shown in Fig. 1. This cell is fabricated with CZ N-type wafer of thickness ~250 μm. The emitter (doped) layer, passivating intrinsic layers and the doped BSF layer of the cell are all thin films (a-Si:H) and deposited by the PECVD technique at ~200°C. The device terminates with a TCO anti reflection coating followed by metallic electrodes.

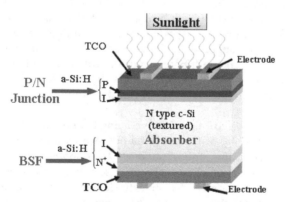

Fig. 1. Schematic diagram of HIT cell proposed by SANYO

HIT cells have (1) potential for high efficiency, (2) very good surface passivation: low surface recombination velocity, (3) low processing temperature - all processes occur at ~ 200°C resulting in low thermal budget, (4) reduced material cost (low temperature processing permits the use of thinner wafers), leading to overall cost reduction and (5) excellent stability- since the base material of the structure continues to be c-Si. With nearly 19 years of steady progress, in 2009, the best HIT solar cells have recorded a efficiency of 23% over a 100.4 cm² cell area (press release SANYO, 2009). Another advantage of HIT solar cells is that it has excellent temperature dependence characteristics and its efficiency does not deteriorate as much as that of diffused junction c-Si cells at higher temperatures (Sakata et al, 2000). The efficiency of HIT cells deteriorates by 0.33%/° C with increase of temperature while it is 0.45%/° C for conventional c-Si solar cells. This means HIT cells would generate more output power in summer time than its diffused junction counterpart.

References	Wafer		Solar cell output parameters				Emitter & BSF deposition technique
	Type	Surface	J_{sc} mA cm^{-2}	V_{oc} mV	Fill factor	η %	
SANYO press release	N	Textured	39.50	729	0.800	23	PECVD
Schmidt et al, 2007	N	Textured	39.3	639	0.789	19.8	RF-PECVD
	P		34.3	629	0.79	17.4	
Wang et al, 2010,2008	P	Textured	36.20	678	0.786	19.3	HWCVD
	N		35.30	664	0.745	17.2	
Olibet et al, 2010, 2007	N	Flat	34.0	680	0.82	19.1	VHF-PECVD
	P		32	690	0.74	16.3	
Das et al, 2008	N	Textured	35.68	694	0.741	18.4	PECVD
Sritharathikhun et al 2008	N	Textured	35.20	671	0.76	17.9	VHF-PECVD
Damon-Lacoste, 2008	P	Flat	33.0	664	0.778	17.1	PECVD
Fujiwara & Kondo, 2009	N	Flat	32.79	631	0.764	17.5	PECVD

Table 1. Summary of best perfoemances of HIT solar cells on P- and N-type c-Si wafer.

Inspired by the outstanding performance of Sanyo HIT cells, many research groups throughout the world have been working with these cells and a-Si:H layers have been deposited by PECVD, hot-wire CVD (HWCVD) and very-high-frequency PECVD (VHF-PECVD). A summary of the best HIT solar cells reported till date is given in Table 1. We find that currently, no group has been able to duplicate what Sanyo has achieved in terms of cell efficiency. Very few groups have reached beyond 19% efficiency: Helmholtz Zentrum Berlin on N-type textured wafers (Schimdt et al, 2007) and the National Renewable Energy Laboratory (NREL) on P-type textured wafers (Wang et al, 2008, 2010) have achieved this feat. Good results have also been obtained by the group of EPFL, IMT, Neuchâtel, Switzrland with high open-circuit voltsge (V_{oc}) on flat wafers. The P-type HIT cell of Damon Lacoste et al (2008) from LPICM-Ecole Polytechnique, France also deserves mention. Here the efficiency is limited by the lower short-circuit current density (J_{sc}) characteristic of flat wafers. The difficulty in attaining the Sanyo HIT cell efficiency illustrates that the a-Si:H/c-Si HJ is indeed a very challenging structure to understand. Therefore, over the last decade scientists are using detailed computer modeling to fully understand the structure. In the next section we will briefly review the computer modeling of HIT solar cells. Recently a few groups have started fabricating HIT cells with intrinsic hydrogenated amorphous silicon oxide (I-a-SiO:H) as the buffer layer between crystalline and doped amorphous silicon. Sritharathikhun et al (2008) have achieved 17.9% cell efficiency with P-μc-SiO:H /N-c-Si cell structure and I-a-SiO:H as the buffer layer. A group from AIST (Fujiwara et al, 2009) has reported 17.5% cell efficiency with a similar cell structure.

2.1 Detailed one-dimensional computer modeling of HIT solar cells:-

Pioneering work in detailed electrical modeling of a-Si:H solar cells was done by Hack and Schur (1985). Other notable models in this respect are the model AMPS (McElheny et al, 1988, Arch et al, 1991) by S. J. Fonash's group at the Pennsylvania State University, USA, the model of Guha's group (Guha et al, 1989), the ASDMP program by P. Chatterjee (Chatterjee, 1992, 1994, 1996), the ASPIN program of Smole and Furlan (1992) and the ASA program by von der Linden et al (1992). Regarding detailed electrical-optical models, which include textured surfaces and light-trapping kinetics to some extent, the first global electrical-optical model developed in the world was when ASDMP was integrated (Chatterjee et al, 1996) to a semi-empirical optical model by Leblanc et al (1994). This program also takes account of specular interference effects for cells with flat surfaces. Later the developed AMPS program (D-AMPS – Plà et al, 2003) and the ASA package, developed at the Delft University of Technology (Zeman et al, 2000) also introduced light trapping effects.

Modeling of HIT cells was started by van Cleef et al (1998 a,b) using the AMPS computer code (McElheny et al, 1988), which however does not have a proper built-in optical model; and the derivative of the AMPS program (D-AMPS), where a fairly good optical model has been introduced (Plà et al, 2003). The numerical PC program AFORS-HET (Stangl et al, 2001, Froitzheim et al, 2002) has been developed especially for simulating HIT solar cells. The latter has recently also been extended to include light-trapping effects. The ASA program in its later version (Zeman et al, 2000) models both the electrical and optical properties of HIT cells. The PC-1D program (Basore, 1990, Basore et al, 1997), developed at the University of News South Wales, Australia for modeling textured mono-crystalline silicon solar cells, has also been fairly successful in modeling HIT cells. The program ASDMP by Chatterjee et al (1994,1996), has also been extended to model N-a-Si:H/P-c-Si type front (with a heterojunction only on the emitter side) (Nath et al, 2008) HIT cells and subsequently used to model double hetreojunction solar cells both on N- and P-type substrates (Datta et al, 2008, 2009, Rahmouni et al, 2010).

2.1.1 Simulation model ASDMP

We will discuss this model in a little more detail, since it has been used in all simulations in this chapter. The "Amorphous Semiconductor Device Modeling Program (ASDMP) " (Chatterjee et al, 1996, Palit et al, 1998), originally conceived to model amorphous silicon based devices, has been extended to also model c-Si and "HIT" cells (Nath et al, 2008). This one-dimensional program solves the Poisson's equation and the two carrier continuity equations under steady state conditions for the given device structure, without any simplifying assumptions, and yields the dark and illuminated current density - voltage (J-V), the quantum efficiency (QE) and the photo- and electro-luminescence characteristics of HIT cells. Its electrical part is described in P. Chatterjee (1994, 1996). The gap state model used in these calculations for the amorphous layers, consists of the tail states and the two Gaussian distribution functions to simulate the deep dangling bond states, while in the c-Si part, the tail states absent. The lifetime of the minority carriers inside the N(P) -c-Si wafer may be estimated using the formula:

$$\tau_p \approx \frac{p - p_0}{R} \text{ or } \tau_n \approx \frac{n - n_0}{R}, \tag{1}$$

where $\tau_p(\tau_n)$, $p(n)$ and $p_0(n_0)$ are the minority carrier lifetime, its density under the given experimental conditions (in this case under 100 mW cm^{-2} of AM1.5 light), and at thermodynamic equilibrium respectively; while R is the recombination rate in the c-Si wafer. The lifetime, calculated in this manner, is in general, position-dependent; however over a large region inside the c-Si wafer, away from the edges, it is a constant and it is this value that is taken to be the minority carrier lifetime in the wafer. van Cleef et al (1998a,b) and Kanevce et al (2009) have also used the DOS model to simulate their HIT cells.

The generation term in the continuity equations has been calculated using a semi-empirical model (F. Leblanc et al, 1994) that has been integrated into the ASDMP modeling program (Chatterjee et al, 1996, Palit et al, 1998). Both specular interference effects and diffused reflectance and transmittance due to interface roughness can be taken into account. The complex refractive indices for each layer of the structure, required as input to the modeling program, have been measured by spectroscopic ellipsometry. In all cases studied in this article, experimentally or by modeling, light enters through the transparent conducting oxide (TCO)/emitter window, which is taken as x = 0 on the position and referred to as the front contact. Voltage is also applied at x = 0. The BSF/ metal contact at the back of the c-Si wafer is taken as x = L on the position scale, where L is the total thickness of all the semiconductor layers of the device. This back contact is assumed to be at ground potential.

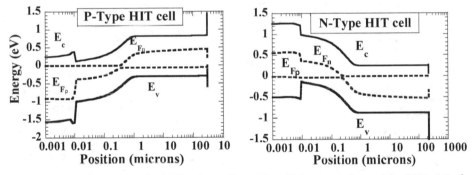

Fig. 2. Energy band diagram for HIT solar cells on P and N type wafers under 100 mW of AM1.5 light and short-circuit conditions.

The calculated energy band diagrams for typical HIT cells on P- and N-type wafers, with passivated surface defects and under 100 mW of AM1.5 light, 0 volts, are shown in Fig. 2.

4. Modeling of HIT solar cells on P-type wafer

4.1 Simulation of experimental results of P-type HIT cells

We have studied both front and double "HIT" structure solar cells on P-type c-Si wafers. These have the structure: N-a-Si:H emitter/ P-c-Si/ aluminum diffused BSF (front HIT) and N-a-Si:H emitter/ P-c-Si/ P+-a-Si:H BSF (double HIT). The experimental data were obtained from the Laboratoire de Physique des Interfaces et des Couches Minces (LPICM), Ecole Polytechnique, Palaiseau, France. Table 2 compares our modeling results to the measured output parameters for front and double HIT structures. Two thicknesses of the N-a-Si:H layer are employed for the front HIT structures, while results are given for two types of double HIT cells having the following structures: (A) 8 nm N-a-Si:H/ 3 nm pm-Si:H intrinsic layer/ P-c-Si wafer/ 23 nm P+-a-Si:H/ 1.5 μm Al, and (B) the above structure, but with a 4 nm P+-a-SiC:H layer sandwiched between the P-c-Si wafer and a 19 nm P+-a-Si:H layer. The pm-Si:H intrinsic layer on the front surface (FS) of the c-Si wafer is there to passivate the defects on this surface. However, no such passivating layer has been deposited on the rear surface (RS) of the c-Si wafer. The defect density on FS was deduced by modeling to be 10^{11} cm^{-2}. Cell B, which has a 4 nm P-type a-SiC:H layer on the rear c-Si wafer surface, has a slightly higher V_{oc} but a lower FF relative to case A, leading to a better efficiency. However, we could not replicate these results in our modeling calculations by the introduction of a P+-a-SiC:H layer of the given properties alone (case B1 in Table 2). In fact, the defect density on the rear wafer surface had to be slightly reduced (case B2, Table 2) to match the experimental results.

Table 2 indicates good agreement between experiments and modeling, except that our modeling results appear to overestimate the FF and hence the efficiency of front HIT cells. In reality this is because screen-printed contacts with low temperature silver paint was used for these cells; resulting in high series resistance and low FF experimentally, which cannot be accounted for by modeling. For double HIT structures, developed later, improved contact formation resulted in very low series resistance and high fill factors experimentally, which agree well with model calculations (Table 2).

HIT type	Sample	N-a-Si:H (nm)	N_{ss} on the DL (cm^{-2})	V_{oc} (mV)	J_{sc} (mA cm^{-2})	FF	η(%)
Front	X1 (E)	12		634	31.90	0.711	14.38
	X1 (M)	12	FS- 4×10^{11}	636	31.85	0.823	16.67
	X2 (E)	8		640	32.54	0.730	15.20
	X2 (M)	8	FS- 4×10^{11}	640	32.57	0.824	17.18
Double	A (E)	8		650	32.90	0.790	16.90
	A (M)	8	FS-10^{11} RS-8×10^{11}	660	32.84	0.781	16.93
	B (E)	8		664	33.10	0.779	17.12
	B1(M)		FS-10^{11} RS-8×10^{11}	653	33.17	0.749	16.24
	B2 (M)	8	FS-10^{11} RS- 3×10^{11}	667	33.21	0.773	17.12

Table 2. Comparison between measured (E) and modeled (M) solar cell output parameters of front and double P-c-Si HIT cells with a flat ITO front contact. DL refers to the defective layer on the wafer surface.

In Fig. 3 (a), we compare the experimentally measured external and internal quantum efficiency (EQE and IQE respectively) curves of the solar cell B to modeling results, while in Fig. 3 (b) we compare the measured IQE curves of a front HIT and the above-mentioned double HIT solar cells, both deposited in the same reactor and under approximately the same conditions of RF power and pressure as solar cells A and B above. The IQE is obtained from the EQE using the formula:

$$IQE(\lambda) = EQE(\lambda) / (1 - R(\lambda) - ITOabs(\lambda)) , \qquad (2)$$

where $R(\lambda)$ is the reflectivity of the HIT cell and $ITOabs(\lambda)$ is the fraction of the light that is absorbed in the transparent conducting oxide, that is indium tin oxide (ITO) in this case.

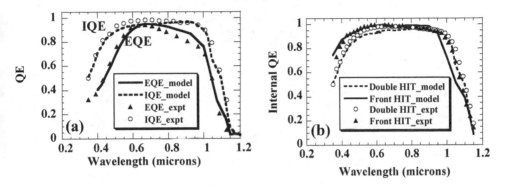

Fig. 3. Comparison of the experimentally measured external and internal QE curves of (a) a double heterojunction cell (case B2 of Table 2); and (b) of a front and the above double HIT solar cell, to modeling results, indicating a higher long wavelength IQE for the double HIT case, both experimentally and in the modeling calculations. The ITO layer is different for the two cases resulting in the difference in the short wave length QE. The lines represent the calculated results, experimental measurements are shown as symbols.

We have used the above simulations to extract the parameters that characterize different layers of the double HIT cells A and B on P-type wafers. These are given in Table 3, together with the extracted parameters of double HIT cells on N-type wafers. The experimental results used to extract the latter and comments thereon, will be discussed in section 5.1. The data in Table 3 includes some measured data: the thickness and doping density of each layer/ wafer, the band gaps of the layers and the electron and hole mobility in the c-Si wafer (Sze, 1981). We also found that a higher value of N_{ss} (as indicated in Case B2 in Table 2 and Table 3) was necessary at the RS to simulate the experimental results. No layer was intentionally deposited to passivate these defects in cells A and B.

Since the 4-nm P+-a-SiC:H layer on the RS of the c-Si wafer (part of the highly doped thin film BSF layer) produces a small but reproducible improvement in the overall device performance, we have tried to understand the basic reasons for this improvement. To realize the role of the thin P-a-SiC:H layer on the RS in case B2 we have made the P+-a-SiC:H layer thicker than in case B2 and adjusted the thickness of following P+-a-Si:H layer to yield a total BSF thickness of 23 nm. We found an all-round deterioration of the solar cell output for the thicker P+-a-SiC:H layers, including a striking fall in the fill factor. We have thus concluded that the introduction of the thin carbide layer as such is not responsible for the observed improvement in cell efficiency of case B2 relative to case A (Table 2). Rather, it appears likely that this wider band gap material helps in passivating the defects on the RS of the c-Si wafer (for which a very thin layer is sufficient) and thereby improves cell performance. In the next section we will discuss how solar cell performance is affected by the defects on the FS and RS of the c-Si wafer.

Parameters	N-a-Si:H/P-a-Si:H emitter	I-pm Si:H buffer	I-a-Si:H buffer	DL on P-c-Si/ N-c-Si on emitter side	P-c-Si/N-c-si wafer	P+-a-Si :H / N+-a-Si : H BSF
Layer thickness (μm)	0.008/0.0065	0.003	0.003	0.003	300/220	0.019
Electron affinity (eV)	4	3.95	4	4.22	4.22	4
Mobility gap (eV)	1.80	1.96	1.80	1.12	1.12	1.78/1.80
Don (accep)doping (cm^{-3})	10^{19}/ 1.41×10^{19}	0	0	9×10^{14}	9×10^{14}	1.4×10^{19}/ 1.45×10^{19}
Eff. DOS in CB (cm^{-3})	2×10^{20}	2×10^{20}	2×10^{20}	2.8×10^{19}	2.8×10^{19}	2×10^{20}
Eff. DOS in VB (cm^{-3})	2×10^{20}	2×10^{20}	2×10^{20}	1.04×10^{19}	1.04×10^{19}	2×10^{20}
Exp.tail prefact. -cm^{-3}eV^{-1}	4×10^{21}	4×10^{21}	4×10^{21}	—	—	4×10^{21}
Charac.energy – VB tail (ED) (eV)	0.05	0.05	0.07	—	—	0.05
Charac.energy – CB tail (EA) (eV)	0.03	0.03	0.04	—	—	0.03
Elec.mobility (cm^2/V-s)	20/25	30	25	1000/1500	1000/1500	20
Hole mobility (cm^2/V-s)	6/5	12	5	450/500	450/500	6/4
Gaussian defect density (cm^{-3})	9×10^{18}	7×10^{14}	9×10^{16}	2.6×10^{18}/ 4.5×10^{18}	10^{12}	8×10^{18}/ 9×10^{18}

Table 3. Input parameters, extracted by modeling, that characterize the above HIT cells. The defect density of 3.3×10^{17} cm^{-3} on the front wafer surface corresponds to a defect density of 10^{11} cm^{-2} (FS) and 3.5×10^{18} cm^{-3} to 8×10^{11} cm^{-2} on the rear surface (RS). The P+-a-SiC:H BSF layer in P-type HIT cells has a larger band gap (1.84 eV), and broader band tails: ED=0.7 eV, EA=0.5 eV

4.2 Influence of the defect density on the front surface of the c-Si wafer:

The effect on the solar cell output parameters of varying the defect density, N_{ss}, on front surface of the P-type c-Si wafer (that which faces the incoming light) is shown in Table 4, using as the base case the double HIT cell B2, but with an assumed textured wafer to reproduce state-of-the-art currents obtainable in HIT cells. The defect density on the RS is held at 10^{11} cm^{-2} for all cases. The results indicate a sharp fall in V_{oc}, and FF.

To understand the sensitivity, we turn to Fig. 4. We note that the electric field is higher at the amorphous - crystalline interface, when $N_{ss} = 3 \times 10^{13}$ cm^{-2} than when $N_{ss} = 10^{11}$ cm^{-2} (Fig. 4a). This is because when the N-a-Si:H layer is joined to a P-c-Si wafer, with a high defect density on its surface, most of the electrons that flow from the N-side to the P-side during junction formation, to bring the thermodynamic equilibrium Fermi levels on either side to the same level, are trapped in these states. The space charge region on the P-c-Si wafer side is therefore localized near the surface and does not extend appreciably into the c-Si wafer. We therefore have a huge density of trapped electrons, a very high interface field (Fig. 4a),

N_{ss} on FS (cm^{-2})	J_{sc} (mA cm^{-2})	V_{oc} (mV)	FF	η(%)
10^{11}	37.50	672	0.770	19.40
2×10^{12}	38.33	586	0.658	14.79
3×10^{13}	38.14	463	0.545	9.65

Table 4. Calculated values of the solar cell output parameters J_{sc}, V_{oc}, FF and η, for different values of the defect density (N_{ss}) on that (front) surface of the crystalline silicon wafer through which light enters, indicating high sensitivity to the V_{oc} and FF. The defect density at the rear surface of the c-Si wafer is 10^{11} cm^{-2}.

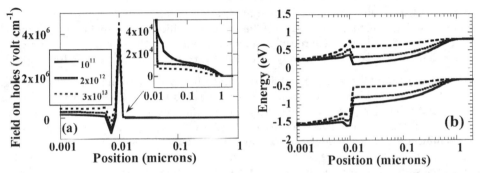

Fig. 4. Effect of changing the defect density (shown in units of cm^{-2}) on the front surface of the c-Si wafer under 100 mW cm^{-2} of AM1.5 light and 0 volts, on (a) the electric field (the inset shows the electric field on an expanded scale over the depletion region) and (b) the band diagram over the front part of the device.

and a collapse of the field over the adjacent depletion region of the c-Si wafer (Fig. 4a inset) for the case with N_{ss} = 3×10^{13} cm^{-2}. This results in the flattening of the energy bands in the totality of the P-type crystalline silicon wafer (Fig. 4b, dashed lines), and a consequent fall in V_{oc} and FF (Table 4). For the case of low N_{ss}, the space charge region on the P-c-Si wafer is not localized and more field exists up to the neutral zone of the c-Si wafer (Fig. 4a inset and band diagram in Fig. 5b, solid lines); resulting in higher V_{oc} and FF (Table 4).

4.3 Influence of the defect density on the rear surface of the c-Si wafer

Table 5 gives the calculated solar cell output parameters J_{sc}, V_{oc}, FF and efficiency for different values of the defect density (N_{ss}) on the rear surface of the c-Si wafer (away from the side where light enters). We have again varied N_{ss} between 10^{11} cm^{-2} and 3×10^{13} cm^{-2}, but this time the largest effect is on the fill factor and the short-circuit current density, as seen from Table 5 and Fig. 5 (a). In order to understand why, we have traced the band diagrams for different N_{ss} on the RS, with the N_{ss} at the FS held at 10^{11} cm^{-2} (Fig. 5b). We find that the band bending over the depletion region has completely disappeared for the highest value of N_{ss} (3×10^{13} cm^{-2}) at RS. From our modeling calculations we also note that up to a defect density of ~10^{12} cm^{-2} at RS, the solar cell output parameters do not deteriorate appreciably. For higher values of N_{ss} the decrease in J_{sc} and FF in particular, is extremely rapid, the sensitivity to V_{oc} being relatively small. Experimentally also it has been found that whether or not an intrinsic passivating layer is deposited on the rear face of the P-type c-Si wafer, the solar cell output is little affected. From this we may conclude that the defect density on the back wafer surface in the experimental as-deposited condition is probably $\leq 10^{12}$ cm^{-2}, as also obtained by modeling the experimental characteristics (Table 2).

N_{ss} on RS (cm^{-2})	J_{sc} (mA cm^{-2})	V_{oc} (mV)	FF	$\eta(\%)$
10^{11}	37.50	672	0.770	19.40
10^{12}	37.48	662	0.752	18.66
2×10^{12}	37.34	625	0.666	15.54
3×10^{13}	5.47	572	0.156	0.49

Table 5. Calculated values of the solar cell output parameters for different values of the defect density (N_{ss}) on that (rear) surface of the crystalline silicon wafer that is away from the incoming light, indicating that the maximum sensitivity is to the short circuit current density and fill factor. The defect density at the front surface of the c-Si wafer is 10^{11} cm^{-2}.

In order to understand the sensitivity of the solar cell output to N_{ss} on the RS, we turn to Fig. 6. We note that when N_{ss} on the rear c-Si wafer surface is highest (3×10^{13} cm^{-2}), there is a huge concentration of trapped holes at the crystalline- amorphous interface on the c-Si wafer side where the high surface defect density exists (dashed line, Fig. 6a). The hole pile-up at the crystalline-amorphous interface slows down the arrival of holes to the back contact (the collector of holes), and encourages the back diffusion of photo-generated electrons in the absorber c-Si wafer. The result is that the electron current is negative over most of the device (Fig. 6b – electron current towards the back contact is negative according to our sign convention). Thus little electron current is collected at the front contact (the collector of electrons, Fig. 6b). In addition, the back-diffusing electrons recombine with the photo-generated holes over most of the absorber c-Si, resulting in poor hole current collection at the back contact. Thus J_{sc} and FF fall sharply for very high values of N_{ss} at RS. More details can be found in Datta et al (2008).

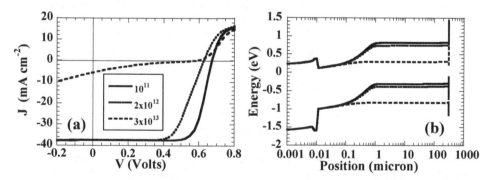

Fig. 5. Effect of changing the defect density (shown in units of cm^{-2}) on the rear surface of the c-Si wafer on (a) the illuminated current density versus voltage characteristics and (b) the band diagram at 0 volts as a function of position in the device under 100 mW cm^{-2} AM1.5 light. Results are shown for double heterojunction solar cells having a 4 nm P$^+$-a-SiC:H/ 19 nm P$^+$-a-Si:H BSF structure. The defect density on the front surface is 10^{11} cm^{-2} for all cases.

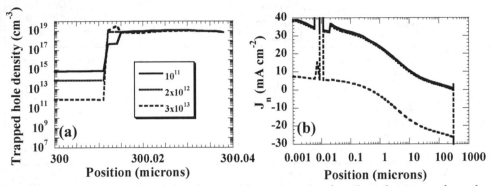

Fig. 6. Effect of changing the defect density (shown in units of cm^{-2}) on the rear surface of the c-Si wafer on (a) the trapped hole density, (b) the electron current density, J_n, and (c) the electric field on the holes. Results are shown for 100 mW cm^{-2} of AM1.5 light under short-circuit conditions.

5. Simulation of N-type HIT solar cells

5.1 Simulation of experimental results

Simulation of a range of experimental results on HIT cells developed by the Sanyo group and available in the literature (Maruyama et al, 2006, Takahama et al, 1992, Sawada et al, 1994, Taguchi et al, 2008) has been undertaken to extract typical parameters that characterize state-of-the-art HIT cells on N-type c-Si substrates, as well as to gain an insight into carrier transport and the general functioning of these cells. Both "front" HIT cells having an amorphous/ crystalline heterojunction on the emitter side only - where the light enters (Takahama et al, 1992), and "double" HIT cells having heterojunctions on both ends of the c-Si wafer (Maruyama et al, 2006, Sawada et al, 1994, Taguchi et al, 2008) have been simulated. The cells have the structure: ITO/ P-a-Si:H/ I-a-Si:H/ textured N-c-Si/ N-c-Si BSF/ metal (front HIT) (Takahama et al, 1992) and ITO/ P-a-Si:H/ I-a-Si:H/ textured N-c-Si/ I-a-Si:H/ N++-a-Si:H/ metal (double HIT) (Maruyama et al, 2006, Sawada et al, 1994, Taguchi et al, 2008). In Taguchi et al (2008), after depositing the undoped and doped a-Si:H layers on both ends of the c-Si wafer, ITO films were sputtered on both sides, followed by screen-printed silver grid electrodes. Simulation of these cells (Maruyama et al, 2006, Takahama et al, 1992, Sawada et al, 1994) gives us an insight into the parameters that play a crucial role in improving HIT cell performance. On the other hand, the article by Taguchi et al (2008) gives the temperature dependence of the dark current density - voltage characteristics and the solar cell output parameters as a function of the thickness of the intrinsic amorphous layer sandwiched between the emitter P-a-Si:H and the main absorber N-c-Si. A study of the temperature dependence of the dark J-V characteristics is particularly important to understand the carrier transport mechanism in these devices. The parameters extracted by such modeling (Table 3) will be used in the following sections to calculate the sensitivity of the solar cell performance to various controlling factors.

In Table 6 we compare our simulation and experimental results of various HIT cells on N-type c-Si substrates (Takahama et al, 1992, Sawada et al, 1994, Maruyama et al, 2006). Modeling indicates that improvements in V_{oc} could be brought about (a) by going from a

HIT	Reference		N_{ss} (cm^{-2}) in defective layers	τ ms	V_{oc} mV	J_{sc} mA cm^{-2}	FF	η %
F	Takahama et al, 1992	E	— —	—	638	37.90	0.775	18.74
		M	FS- 4x10^{11}	0.23	643	37.89	0.775	18.88
D-I	Swada et al, 1994	E	— —	—	644	39.40	0.790	20.05
		M	FS-4x10^{11} RS -10^{11}	0.5	658	39.03	0.783	20.11
D-II	Maruyama et al, 2006	E	— —	1.20	718	38.52	0.790	21.85
		M	FS & RS - 10^{11}	2.00	713	38.60	0.797	21.93

Table 6. Comparison between measured (E): and modeled (M) solar cell output of front (F) and double (D) N-c-Si HIT cells with textured ITO front contact, developed by Sanyo over the years. "τ" is the lifetime of the minority carriers in the c-Si wafer.

front HIT to a double HIT structure, (b) by decreasing the defects on the front surface of the c-Si wafer that faces the emitter layer and (c) by improving the lifetime of the minority carriers in crystalline silicon. Results indicate that it is by decreasing N_{ss} on the front surface of the c-Si wafer, that the largest increase in V_{oc} could be achieved, without any fall in FF.

We next used ASDMP to simulate the experimental results of Taguchi et al (2008). Here we have concentrated on the effect of varying the thickness of the intrinsic amorphous silicon layer at the P-amorphous emitter/ N-c-Si heterojunction. The terminology "normal" has been used to represent the thickness of the front I-a-Si:H buffer layer in the cell that yields the highest efficiency (Table 7). Modeling reveals that the I-a-Si:H thickness for this case is 3 nm. The I-a-Si:H buffer layers (front) in the cells named "Half", "Double" and "Triple" by Taguchi et al (2008) have therefore been assigned thicknesses of 1.5 nm, 6 nm and 9 nm respectively in the simulations. Results of our simulation of the experimental light J-V characteristics (Taguchi et al, 2008) as a function of this I-a-Si:H layer thickness are given in Table 7 and the input parameters extracted by such modeling, and also of the dark J-V characteristics (Figs. 7a and 7b) and typical internal quantum efficiencies of Sanyo N-c-Si HIT cells (Fig. 7c, Maruyama et al, 2006), are given in Table 3 (the same table that contains the extracted parameters of P-type HIT cells). Since modeling does not consider the resistance of the contacts; these results had to be modified by taking into account the series resistance of the contacts. The addition of the series resistance did not modify V_{oc} and J_{sc} but allowed to perfectly match the experimental fill factor and therefore the efficiency of the Sanyo HIT solar cells (Taguchi et al, 2008). In Table 7 we show the solar cell output parameters as obtained directly by modeling, without resistive losses (which gives an upper limit for the FF and therefore the efficiency) and the values of the FF and efficiency after considering the constant series resistance (marked by asterisks). This resistance, comprising resistive losses in the TCO, the silver grid and the contacts, was estimated by Taguchi et al (2008) to be ~2.8 mΩ.

Cell name	μ_n (μ_p) cm²/volt-sec	I-a-Si:H thickness(nm)	N_{ss} (cm⁻²)		J_{sc} (mA cm⁻²)	V_{oc} (volts)	FF	η(%)
Half	30 (6)	1.5	4x10¹¹	E	37.4	0.699	0.776	20.3
				M	37.2	0.702	0.803 0.775*	21.0 20.2*
Normal	25 (5)	3.0	1.5x10¹¹	E	37.2	0.711	0.773	20.4
				M	37.0	0.712	0.799 0.774*	21.0 20.4*
Double	15 (3)	6.0	10¹⁰	E	36.5	0.718	0.747	19.6
				M	36.7	0.717	0.766 0.747*	20.2 19.7*
Triple	15 (3)	9.0	10¹⁰	E	36.4	0.715	0.717	18.7
				M	36.6	0.714	0.750 0.718*	19.6 18.8*

Table 7. Modeling (M) of the experimental (E) results of N-type HIT solar cells, having different thickness of the I-a-Si:H layer on the emitter side. N_{ss} is the defect density on that surface of the c-Si wafer that faces the emitter. The quantities given with astericks are the calculated values of FF and efficiency corrected for the series resistance of the contacts (2.8 m mΩ) (Taguchi et al, 2008).

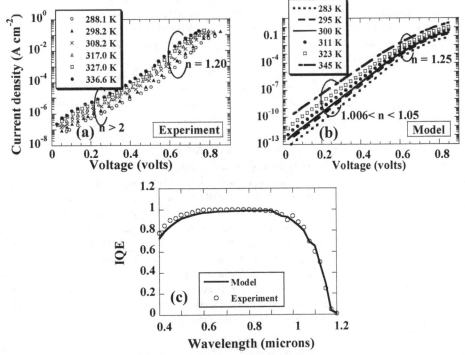

Fig. 7. (a) Experimental (Taguchi et al, 2008) and (b) simulated dark J-V characteristics of the cell with "Normal" thickness I-a-Si:H layer at the P-a-Si:H/ N-c-Si interface at various temperatures, and (c) the IQE of the same cell under AM1.5 illumination and 0 volts compared to the experimental IQE of a typical Sanyo cell (Maruyama et al,2004).

The dangling bond defect density in the I-a-Si:H layer, as extracted from modeling, is 9×10^{16} cm^{-3} and its Urbach enegy is 70 meV (Table 8). We have assumed the same values for these quantities, as well as of the capture cross-sections of the defect states inside the I-a-Si:H layer in all the cases of Table 7. Modeling indicates that in order to simulate the lower V_{oc}'s of the Taguchi et al (2008) cells "Normal" and "Half", the defect density on the surface of the c-Si wafer itself in these cases, must be higher (Table 7). We may justify this fact by assuming that a very thin buffer layer may not be as effective in passivating the defects on the surface of the c-Si wafer as a thicker buffer layer. In Table 7, we also had to assume higher carrier motilities in the front amorphous layers for the cases Half and Normal to match both the higher FF and lower V_{oc} for these cases. Increasing carrier mobilities over the front amorphous layers improves hole collection and therefore the FF. However, higher electron mobility allows more electrons to recto-diffuse towards the front contact (collector of photo-generated holes) and recombine with holes, thus reducing V_{oc}. However the main reason for the lower V_{oc} for thinner I-a-Si:H layers (Half and Normal) is our assumption of higher surface defect density on the c-Si wafer in these cases (Table 7).

The experimental dark J-V characteristics of the cell "Normal" is shown in Fig. 7 (a) and the model curves in Fig. 7 (b). The diode ideality factor, n, calculated in the voltage range 0.4 volts$\leq$ V< 0.8 volts, from the model dark characteristics is 1.25 and compares well with the experimental value of 1.2. This value of "n" indicates that it is the diffusion current that dominates transport in this voltage range for N-c-Si HIT cells, as is also the case for homojunction c-Si solar cells. On the other hand, in the voltage range 0.1 volts< V< 0.4 volts, "n" calculated from the modeling data (Fig. 7b) is ~1, which indicates that the conductivity continues to be dominated by diffusion. The value of the slope, calculated from the experimental curves of Taguchi et al (2008) in the voltage range 0.1 volts < V < 0.4 is smaller than that of the recombination current model and remained almost constant for each temperature. The corresponding value of "n" derived from the experimental curves is greater than 2 (Fig. 7a). Taguchi et al (2008) therefore assumed that this is tunneling-limited current. If the value of "n" extracted from the experimental curves, had been due to current dominated by recombination, ASDMP would also have been able to reproduce this value of 'n', since the recombination current model is included in ASDMP. In fact ASDMP has already been used to successfully model forward and dark reverse bias characteristics of a-Si:H based PIN solar cells, where recombination plays a dominant role (Tchakarov et al, 2003). The fact that the value of "n" calculated from the ASDMP-generated dark J-V curves is ~1, while that from experiments is different, indicates that the current over this region is dominated by a phenomenon *not* taken account of by ASDMP (e.g. tunneling). Over this voltage region therefore the current could be dominated by the tunneling of electrons. However, as pointed out by Taguchi et al (2008), "the current density in this region is sufficiently low compared to the levels of short-circuit current density and does not affect solar cell performance". It therefore appears that cell performance under AM1 or AM1.5 light is not affected by tunnelling of electrons, although this phenomenon probably exists for V< 0.4 volts.

Fig. 8 shows the temperature dependence of the solar cell output parameters. We have made the comparison between experiments (Taguchi et al, 2008) and modeling, after taking account of the series resistance of the contacts that is independent of temperature. As the temperature decreases, carrier density decreases. It means less carrier recombination and therefore a higher V_{oc} at lower temperatures (Fig. 8a). However lower carrier density at

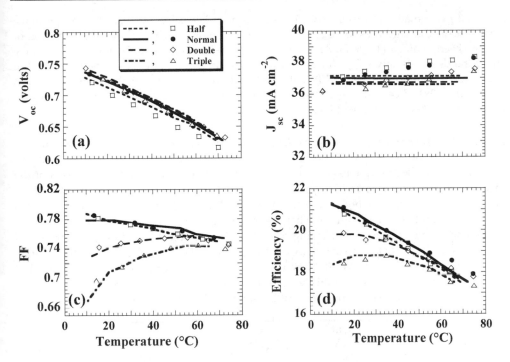

Fig. 8. Variation of (a) V_{oc}, (b) J_{sc}, (c) FF and (d) Efficiency as a function of temperature in N-c-Si HIT solar cells having different thickness of the undoped a-Si:H layer (half, normal, double, triple) at the P-a-Si:H/ N-c-Si interface. The lines are modeling results, while symbols correspond to measured data.

lower temperatures, also means that the cell is now more resistive, resulting in a fall in the FF for the cells "double" and "triple" (Fig. 8c), where performance is dominated by the undoped a-Si:H layer. Also, for the value of the band gap assumed for the I-a-Si:H layer (Table 8), the holes are able to overcome the positive field barrier at the a-Si/ c-Si interface by thermionic emission to get collected at the front contact. Thermionic emission decreases at lower temperatures, resulting in a loss of FF for cells "double" and "triple". For cells "Normal" and "Half", performance is dominated by the temperature-independent resistance of the contacts; therefore no fall in FF is seen. Finally Fig. 8 (b), indicates that the calculated J_{sc} is constant with temperature, while the measured J_{sc} increases slightly. This is because the model does not take account of the temperature dependence of the band gap and absorption coefficient of the materials.

5.2 Effect of I-a-Si:H buffer layers on the performance of N- type HIT solar cells
HIT solar cells give efficiencies comparable to those of c-Si cells because of the amazing passivating properties of the intrinsic a-Si:H layers. In fact it is this layer that gives this group of solar cells its name – "HIT". We have already discussed that it is very effective in passivating the defects on the surface the c-Si wafer. However, it must be kept as thin as possible, as it reduces the fill factor when thick (Table 7). We have next studied the effect on

solar cell performance of varying the defect density in this layer itself. For this purpose, we have assumed its thickness to be 6 nm (as in case "Double") where the best passivation of N_{ss} has been attained (Table 7). An increase in the defect density in the I-a-Si:H layer may affect the defect density (N_{ss}) on c-Si, but in this study we assume N_{ss} to be constant. We have found (Rahmouni et al, 2010) that unless the defect density of this intrinsic layer is greater than $3x10^{17}$ cm^{-3}, no significant loss of cell performance occurs. Similar conclusions have been reached in the case of HIT cells on P-type c-Si wafers.

5.3 Effect of the defect density on the front and rear faces of the N-type c-Si wafer

The sensitivity of the solar cell output of HIT cells on N-type wafers to the surface defect density (N_{ss}) at the amorphous/crystalline interface is given in Table 9. All aspects of the solar cell output appear to be highly sensitive to the N_{ss} on the front surface (on the side of the emitter layer) of the N-type c-Si wafer; however the sensitivity to N_{ss} on the rear face is weak and is limited to the condition when these defects are very high. We have also given in Table 8, the values of the corresponding recombination speeds at the a-Si:H /c-Si front and the c-Si/a-Si:H rear heterojunctions, as calculated by ASDMP, under AM1.5 illumination and short circuit condition. We find that for a well-passivated front interface ($N_{ss} \leq \sim 3x10^{11}$ cm^{-2}) the recombination speed at this heterojunction is less than 10 cm/sec (Table 8), in good agreement with measured interface recombination speeds (Dauwe et al, 2002).

N_{ss} at front (DL) (cm^{-2})	S_p at front (DL) (cm/s)	N_{ss} at back (DL) (cm^{-2})	S_n at back (DL) (cm/s)	Jsc (mA cm^{-2})	V_{oc} (volts)	FF	η(%)
10^{10}	3.62			36.96	0.720	0.801	21.32
$1.5x10^{11}$	4.20			37.00	0.712	0.799	21.03
10^{12}	24.73	10^{10}	$2.89x10^4$	37.24	0.636	0.695	16.46
$2x10^{12}$	202.62			37.37	0.596	0.470	10.47
10^{13}	$1.16x10^3$			18.83	0.544	0.160	1.64
$1.5x10^{11}$	4.20	10^{10}	$2.89x10^4$	37.00	0.712	0.799	21.03
		10^{11}	$2.37x10^4$	36.99	0.711	0.799	21.01
		10^{12}	$1.95x10^4$	36.98	0.696	0.797	20.51
		10^{13}	$1.00x10^4$	35.45	0.609	0.779	16.82

Table 8. Sensitivity of the solar cell output to the defect density (N_{ss}) in thin surface layers (DL) on the front and rear faces of the c-Si wafer in N type double HIT solar cells. The P-layer thickness is 6.5 nm. The recombination speeds of holes (S_p - at the front DL) and electrons (S_n - at the rear DL), calculated under AM 1.5 light and 0 volts, are also shown.

In Fig.9 (a) we plot the light J-V characteristics and in Fig. 9 (b) the band diagram for various values of N_{ss} on the front face of the c-Si wafer. We find that for a very high defect density on the surface of the c-Si wafer, the depletion region in the N-c-Si wafer completely vanishes, while the emitter P-layer is depleted (Fig. 9b). With a high N_{ss} on the c-Si wafer, the holes left behind by the electrons flowing into the P-layer during junction formation, are localized on its surface, leading to a high negative field on the wafer surface and little field penetration into its bulk (Fig. 10a). Hence the near absence of the depletion zone in N-c-Si and a strong fall in V_{oc} for the highest N_{ss} (10^{13} cm^{-2}).

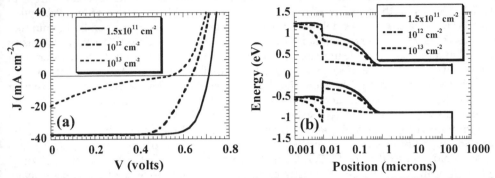

Fig. 9. (a) The light J-V characteristics and (b) the band diagram under AM1.5 light bias and 0 volts for different values of N_{ss} on the front face of the N type c-Si wafer.

In Fig. 10 (b) we plot the trapped hole population over the front part in N-c-Si double HIT cells under AM1.5 bias light at 0 volts. We note that when N_{ss} on the front c-Si wafer surface is the highest (10^{13} cm^{-2}), there is a huge concentration of holes at the amorphous / crystalline (a-c) interface on the c-Si wafer side, where the high surface defect density exists (dashed line, Fig. 10b).

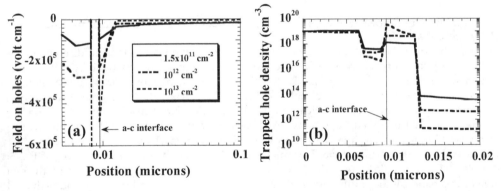

Fig. 10. Plots of (a) the electric field on the holes and (b) the trapped hole density over the front part of the device as a function of position in the entire device under illumination and short-circuit conditions, in N-c-Si HIT cells for different densities of defects on the front face of the c-Si wafer. The amorphous/crystalline (a-c) interface is indicated on (a) and (b).

The hole pile-up at the amorphous / crystalline interface slows down the arrival of holes to the front contact (the collector of holes), and attracts photo-generated electrons, i.e., encourages their back diffusion towards the front contact. The result is that the electrns back-diffuse towards the front contact and recombine with the photo-generated holes resulting in poor carrier collection (Rahmouni et al, 2010). Thus J_{sc} and FF fall sharply for high values of N_{ss} on the front surface of c-Si (Table 8). In fact we may arrive at the same conclusion also from Fig. 9 (b), which shows that for $N_{ss} = 10^{13}$ cm^{-2}, there is almost no band bending or electric field in the c-Si wafer (the main absorber layer) so that carriers cannot be collected, resulting in the general degradation of all aspects of solar cell performance.

On the other hand Table 8 indicates that there is little sensitivity of the solar cell output to the defect states on the rear face of the wafer, except at the highest value of N_{ss}. To explain this fact, we note that the recombination over the rear region is determined by the number of holes (minority carriers) that can back diffuse to reach the defective layer. Not many succeed in doing so, since the high negative field due to the large valence band discontinuity at the c-Si/ a-Si rear interface pushes the holes in the right direction, in other words, towards the front contact. Therefore the defects over this region cannot serve as efficient channels for recombination, and there is no large difference between the recombination through these states for different values of N_{ss} (Table 8). Moreover the conduction band discontinuity at the c-Si/ a-Si interface is about half that of the valence band discontinuity. Since the mobility of electrons, relative to that of holes, is also much higher, clearly this reverse field due to the conduction band discontinuity poses little difficulty for electron collection even when the defect density at this point is high, except when $N_{ss} \geq 10^{13}$ cm^{-2}, from which point the solar cell performance deteriorates.

6. Comparative study of the performances of HIT solar cells on P- and N-type c-Si wafers

Using parameters extracted by our modeling (given in Tables 3), we have made a comparative study between the performances of HIT solar cells on 300 μm thick textured P- and N-type c-Si wafers (for more details refer to Datta et al, 2010).

6.1 Sensitivity of amorphous/crystalline band discontinuity in the performances of HIT solar cells

Since the band gap, activation energy of the amorphous layers and the band discontinuities at the amorphous/crystalline interface are interlinked, we treat these sensitivity calculations together. For HIT cells on P-c-Si, the large valence band discontinuity (ΔE_v) on the emitter side prevents the back-diffusion of holes and has a beneficial effect. Keeping this constant, we varied the mobility gap and therefore the conduction band discontinuity (ΔE_c) on the emitter side. We find that a ΔE_c upto 0.3 eV, does not impede electron collection, but instead brings up both J_{sc} and V_{oc}, due to an improved built in ptential (V_{bi}).

However high ΔE_v at the crystalline/amorphous (c-a) interface on the BSF side of P-c-Si double HIT cells (Table 9), impedes hole collection, resulting in a pile up of holes on the c-Si side of this band discontinuity (Fig. 11a) and a consequent sharp fall in the FF and S-shaped J-V characteristics for high ΔE_v, especially when the activation energy of the P-a-Si:H layer is also high (Fig. 11b).

E_μ (P) (eV)	E_{ac} (eV)	ΔE_v (eV)	J_{sc} (mA cm^{-2})	V_{oc} (mV)	FF	η %
1.75	0.3	0.41	36.70	649	0.810	19.28
1.75	0.4	0.41	36.69	647	0.688	16.34
1.80	0.3	0.46	36.70	649	0.807	19.21
1.90	0.3	0.56	36.70	649	0.762	18.14
1.90	0.4	0.56	36.68	649	0.484	11.51
1.98	0.4	0.64	27.45	649	0.171	3.04

Table 9. Variation of solar cell output with mobility gap (E_μ), activation energy (E_{ac}) and ΔE_v (P-c-Si/P-a-Si:H BSF interface) in double P-c-Si HIT solar cells. ΔE_c is held constant at 0.22eV.

It is for this reason that a transition from a front to a double HIT structure does not appreciably improve cell performance for P-c-Si HIT cells. The accumulated holes at the c-a interface, furthermore, repel the approaching holes and encourage photo-generated electron back diffusion, resulting in increased recombination, that reduces even J_{sc} for the highest ΔE_v (Table 9, Fig. 11b). Finally, for high hole pile-up, the amorphous BSF is screened from the rest of the device, so that the large variation of its band gap and activation energy (Table 9) fails to alter the V_{oc} of the device. The best double HIT performance is attained when the mobility gap (ΔE_μ) of the amorphous BSF P-layer is ≤ 1.80 eV and $E_{ac} = 0.3$ eV (Table 9).

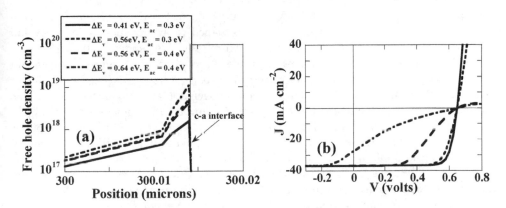

Fig. 11. Variation of (a) the free hole population near the c-Si/ amorphous BSF interface and (b) the light J-V characteristics for different valence band discontinuities (ΔE_v) and activation energies (E_{ac}) of the P-BSF layer in double P-c-Si HIT solar cells. $\Delta E_c = 0.22$eV in all cases.

Table 10 shows the effect of the variation of the emitter P-layer mobility gap, activation energy and the valence band discontinuity at the a-c interface on N-c-Si double HIT cell performance.

E_μ (P) (eV)	E_{ac} (eV)	ΔE_v (eV)	J_{sc} (mA cm^{-2})	V_{oc} (mV)	FF	η (%)
1.75	0.3	0.41	38.06	670	0.818	20.86
1.75	0.4	0.41	38.14	652	0.681	16.93
1.80	0.3	0.46	38.10	671	0.811	20.75
1.90	0.3	0.56	38.22	677	0.705	18.25
1.90	0.4	0.56	38.38	674	0.463	11.98
1.98	0.4	0.64	28.18	732	0.184	3.79

Table 10. Variation of solar cell output parameters with mobility gap (E_μ), activation energy (E_{ac}), and ΔE_v at the emitter P-a-Si:H/c-Si interface in double N-c-Si HIT solar cells. ΔE_c is held constant at 0.22eV.

Table 10 indicates that for valence band offsets up to 0.51 eV, and E_{ac} (P) ≤ 0.3 eV, the FF is high, indicating that the majority of the holes photo-generated inside the c-Si wafer, can surmount the positive field barrier due to the a-Si/ c-Si valence band discontinuity by thermionic emission and get collected at the front ITO/ P-a-Si:H contact. However solar cell performance deteriorates both with increasing band gap and increasing E_{ac} of the P-layer. The latter is only to be expected as it reduces the built-in potential.

Fig. 12 (a) shows the effect on the energy band diagram of increasing the P-layer band gap (therefore of increasing ΔE_v, since ΔE_c is held constant) and the activation energy. Increasing ΔE_v at the P-a-Si:H/N-c-Si interface results in hole accumulation and therefore a fall in FF for $\Delta E_v \geq 0.56$ eV, for a P-layer activation energy of ~0.3 eV, due to the reverse field it generates; that is further accentuated when E_{ac} is high (Table 10). van Cleef et al (1998 a,b) have also shown that for a P-layer doping density of 9×10^{18} cm^{-3} (same as ours – Table 3, giving E_{ac} = 0.3 eV) and for ΔE_v = 0.43 eV, normal J-V characteristics are achieved at room temperature and AM1.5 illumination, and that "S-shaped" characteristics begin to develop at higher ΔE_v and E_{ac}. In our case, for $\Delta E_v \geq 0.60$ eV, Fig. 12(c) indicates that free holes accumulate over the entire c-Si wafer, resulting in a sharp reduction of the electric field and flat bands over the depletion region, on the side of the N-type c-Si wafer (Fig. 12b). This fact results in a sharp fall in the FF and conversion efficiency (Table 10). In fact under this condition, the strong accumulation of holes on c-Si, can partially deplete even the highly defective P-layer, resulting in a shift of the depletion region from c-Si to the amorphous emitter layer (Fig. 12a). This also means that the carriers can no longer be fully extracted at 0 volts, resulting in a fall in J_{sc} (Table 10). We have found that the current recovers to the normal value of ~36 mA cm^{-2} only at a reverse bias of 0.3volts (Datta et al, 2010). Modeling indicates that for improved performance of N-c-Si HIT cells, the valence band offset has to be reduced by a lower emitter band gap, unless the tunneling of holes exists.

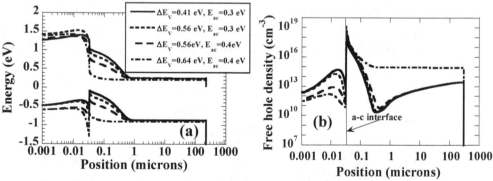

Fig. 12. Variation of (a) the band diagram under AM1.5 light and 0 volts and (b) the free hole population under the same conditions, as a function of position in the N-c-Si HIT device for different valence band discontinuities (ΔE_v) and activation energies (E_{ac}) of the emitter layer.

6.2 Sensitivity of the solar cell output to the front contact barrier height.

The front TCO/P-a-Si:H contact barrier height, ϕ_{b0} in N-type HIT cells is determined by the following expression:-

$$\phi_{b0} = E_\mu(P) - E_{ac}(P) - sbb ,$$

                                                                                            (3)

where $E_\mu(P)$ and $E_{ac}(P)$ represent respectively the mobility band gap and the activation energy of the P-layer, and 'sbb' is the surface band bending due to a Schottky barrier at the TCO/P interface. With a change of the work function of the TCO, it is this 'sbb' that varies. In this section we study the dependence of the solar cell output to changes in this surface band bending. We hold the band gap and the activation energies of the P-layer constant at 1.75 eV and 0.3 eV respectively, so that the TCO work function has a direct effect on the front contact barrier height. The results are summarized in Fig. 15. For these sensitivity calculations we have chosen the thickness of the P-layer to be 15 nm (Rahmouni et al, 2010). Fig. 13 indicates that both V_{oc} and FF fall off for $\Phi_{b0} \leq 1.05$ eV.

We have also studied the effect of changing the rear P-a-Si:H BSF/TCO barrier height , ϕ_{bL} , in P-c-Si HIT cells. The variation in the current-density – voltage characteristics follow a similar pattern as Fig. 15.

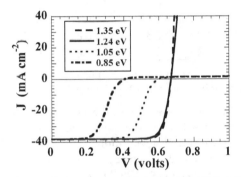

Fig. 13. The current density - voltage characteristics under AM1.5 light and 0 volts for different front contact barrier heights. The band gap, the activation energy and the thickness of the P-layer are held constant at 1.75 eV, 15 nm and 0.3 eV respectively, so that only surface band bending changes.

6.3 Relative influence of different parameters on the performance of HIT cells

In this section we make a comparative study of the influence on HIT cell performance, of the N_{ss} on the surface of the c-Si wafer, the lifetime (τ) of the minority carriers in c-Si, and the surface recombination speeds (SRS) of free carriers at the contacts. The sensitivity to the first two is shown in Table 11. For all the cases studied here, the P layer has an activation energy of 0.3 eV and a surface band bending 0.21 eV.

We note that when the defect density on the surfaces of the c-Si wafer is low, there is some sensitivity of the solar cell output to τ. In fact the conversion efficiency increases by ~3.22% and ~2.47% in double P-c-Si and N-c-Si HIT cells respectively as τ varies from 0.1 ms to 2.5 ms. By contrast there is a huge sensitivity to N_{ss}, as already noted in sections 4.2, 4.3 and 5.3; the performance of the HIT cell depending entirely on this quantity when it is high, with no sensitivity to τ (Table 11). The lone exception is the N_{ss} on the rear face of N-c-Si, to which solar cell output is relatively insensitive as already noted

Finally, the minority carrier SRS at the contacts, that regulates the back diffusion of carriers, has only a small influence in these double HIT cells. The majority carrier SRS does not affect cell performance up to a value of 10^3 cm/s, except the SRS of holes at the contact that is the

Type	N_{ss} (cm^{-2})		τ	J_{sc}	V_{oc}	FF	η
	Front	Rear	(ms)	(mA cm^{-2})	(mV)		(%)
P-c-Si	4x10^{11}	10^{11}	0.1	36.22	604	0.794	17.37
			0.5	36.61	649	0.808	19.19
			2.5	36.68	687	0.817	20.59
	3x10^{13}	10^{11}	0.5	37.24	472	0.626	11.00
			2.5	37.17	471	0.626	10.96
	4x10^{11}	3x10^{13}	0.5	5.68	572	0.154	0.50
			2.5	5.59	572	0.153	0.49
N-c-Si	4x10^{11}	10^{11}	0.1	38.39	631	0.767	18.58
			0.5	39.03	658	0.783	20.13
			2.5	39.20	678	0.792	21.05
	3x10^{13}	10^{11}	0.5	11.54	537	0.208	1.29
			2.5	11.58	537	0.207	1.29
	4x10^{11}	3x10^{13}	0.5	37.04	615	0.763	17.39
			2.5	37.08	616	0.763	17.44

Table 11. Sensitivity of double HIT solar cell output parameters to N_{ss} on the front and rear surfaces of the c-Si wafer and minority carrier life-time (τ).

hole-collector. Hole collection (at the rear contact in P-c-Si HIT and at the front in N-c-Si HIT) is already somewhat impeded by the large valence band discontinuity at the amorphous/ crystalline interface and the lower mobility of holes relative to electrons; hence a low value of SRS of holes at the contacts is expected to have a disastrous influence on hole collection. The effect of lowering S_{p0} for N-c-Si HIT cells is shown in Fig. 14, and is seen to lead to S-shaped J-V characteristics with a sharp fall in the FF when reduced to $\leq 10^4$ cm/sec. In fact when sputtering ITO onto c-Si substrates coated with a-Si:H (intrinsic and doped) films, we sometimes obtain a rather degraded P/ITO interface, where the surface recombination speed is probably reduced. Therefore, Fig. 14 indicates that ITO deposition conditions can also be critical for good solar cell performance.

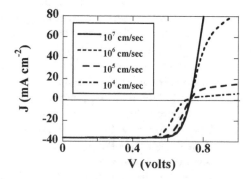

Fig. 14. The sensitivity of the illuminated J-V characteristic under AM1.5 light and short-circuit condition, to the surface recombination speed of the holes at the ITO/P front contact.

7. Conclusions

We have studied the performance of HIT cells on P-and N-type c-Si wafers, using detailed computer modeling. In order to arrive at a realistic set of parameters that characterize these cells, we have modeled several experimental results. We find that the major breakthroughs in improving the performance of these cells having textured N-type c-Si as the absorber layer, come from the introduction of an amorphous BSF layer, by passivating the defects on the c-Si wafer surface and, to a lesser extent, by improving the lifetime of the minority carriers in the c-Si wafer (Table 6).

Modeling indicates that both types of HIT cell output is very sensitive to the defects on the surface of the c-Si wafer, and good passivation of these defects is the key to attaining high efficiency in these structures. An exception to this rule is the defects on the rear face of c-Si in N-type HIT cells, to which there is not much sensitivity. The amorphous/crystalline valence band discontinuity also has a strong impact. In particular, large ΔE_v at the emitter P-a-Si:H/N-c-Si contact leads to S-shaped J-V characteristics, unless tunneling of holes takes place; while that at the P-c-Si/P-BSF contact reduces the FF in double P-c-Si HIT cells. It is for this reason that a transition from a front to double HIT structure on P-c-Si does not produce the spectacular improvement observed for N-type HIT cells (Table 6). Solar cell output is also influenced to some extent by the minority carrier lifetime in c-Si. In Table 12 we compare the performance of a P-type and an N-type HIT cell, with low N_{ss} on the wafer surface, and realistic input parameters. We find that the N-type HIT cell shows better performance than a P-c-Si HIT cell with a higher V_{oc} and conversion efficiency, because of a higher built-in potential in the former. However, the fill factor of N-c-Si HIT cells is lower than in P-type HIT cells due to the assumption of $\Delta E_v > \Delta E_c$, resulting in the holes facing more difficulty in getting collected at the front contact in the former case. This fact has also been pointed out by other workers (Stangl et al, 2001, Froitzheim et al, 2002). In P-type HIT cells, the electrons are collected at the front contact and have to overcome the relatively low ΔE_c at the crystalline/amorphous interface so that its FF is higher than in N-c-Si HIT.

Type	J_{sc} (mA cm^{-2})	V_{oc} (mV)	FF	η (%)
Double HIT on P-c-Si	37.76	694	0.828	21.72
Double HIT on N-c-Si	38.89	701	0.814	22.21

Table 12. Comparison of the performance of P-type and N-type double HIT cells, with optimized parameters. The life time of minority carriers in the c-Si wafer in both cases is 2.5 ms and its doping 10^{16} cm^{-3}.

8. Acknowledgements

The authors wish to express their gratitude to Prof. Pere Roca i Cabarrocas of LPICM, Ecole Polytechnique, Palaiseau, France for providing all the experimental results on "HIT" cells on P-types wafers, that have been simulated in this article. We are also grateful to him for many in-depth discussions and constant encouragement during the course of this work. The authors also wish to thank Prof. C. Baliff, of IMT, University of Neuchâtel, Switzerland, M. Nath of the Energy Research Unit, IACS, Kolkata, India and J. Damon-Lacoste of TOTAL, S. A. for many helpful discussions.

9. References

Arch, J. K.; Rubinelli, F. A.; Hou, J. Y. and Fonash, S. (1991) Computer analysis of the role of p-layer quality, thickness, transport mechanisms, and contact barrier height in the performance of hydrogenated amorphous silicon p-i-n solar cells, *Journal of Applied Physics* Vol.69, No. 10(May, 1991) pp 7057 -7066, ISSN 0021-8979.

Basore, P. A. (1990), Numerical modeling of textured silicon solar cells using PC-1D IEEE Transaction on Electron Devices, Vol 37, No. 2 (February, 1990) pp. 337 – 343, ISSN 0018-9383 .

Clugston, D.A and Basore P. A. (1997), PC1D version 5: 32-bit solar cell modeling on personal computers, *Proceedings of 26th IEEE Photovoltaic Specialists Conference*, pp. 207- 201, ISBN: 0-7803-3767-0, Anaheim, USA, 1997, September 29-October 3.

Chatterjee, P. (1992), Computer modeling of the dependence of the J-V characteristics of a-Si:H solar cells on the front contact barrier height and gap state density, Technical Digest of International PVSEC-6, New Delhi, India, Feb. 10-14 (1992) pp 329-334.

Chatterjee P. (1994), Photovoltaic performance of a-Si:H homojunction p-i-n solar cells: A computer simulation study, *Journal of Applied Physics* , Vol 76 No.2 (July, 1994) pp 1301-1313, ISSN 0021-8979.

Chatterjee, P. (1996), A computer analysis of the effect of a wide-band-gap emitter layer on the performance of a-Si:H-based heterojunction solar cells, *Journal of Applied Physics*, Vol 79, No 9 (May, 1996) pp 7339-7347, ISSN (print) 0021-8979.

Chatterjee, P.; Leblanc, F.;Favre; M. and Perrin, J. (1996), A global electrical-optical model of thin film solar cells on textured substrates, *Mat. Res. Soc. Symp. Proc.* 426 (1996) pp 593-598.

Datta A., Damon-Lacoste J., Roca i Cabarrocas P., Chatterjee P. (2008), Defect states on the surfaces of a P-type c-Si wafer and how they control the performance of a double heterojunction solar cell, *Solar energy Materials and Solar Cells*, Vol 92 (August, 2008) pp 1500-1507, ISSN 0927-0248.

Datta, A.; Damon-Lacoste, J.; Nath, M.; Roca i Cabarrocas, P. andChatterjee P. (2009), Dominant role of interfaces in solar cells with N-a-Si:H/P-c-Si heterojunction with intrinsic thin layer, *Materials Science and Engineering B*, Vol 15-160 (2009), 10-13.

Datta, A.; Rahmouni, M.; Nath, M.; Boubekri, R.; RocaiCabarrocas, P. and Chatterjee, P. (2010) , Insights gained from computer modeling of heterojunction with instrinsic thin layer "HIT" solar cells, *Solar energy Materials and Solar Cells*, Vol 94, (April, 2010) pp 1457-1462, ISSN 0927-0248.

Damon-Lacoste J., Ph. D Thesis, Ecole Polytechnique Paris, 2007.

Dauwe, S; Schmidt, J. and Hezel, R.,2002, Very low surface recombination velocities on p- and n-type silicon wafers passivated with hydrogenated amorphous silicon films, *Proceedings of 29th IEEE Photovoltaic Specialists Conference*, pp.1246 – 1249, ISBN 0-7803-7471-1, New Orleans, USA, 2002, May 19-24.

Froitzheim, A.; Stangl, R.; Elstner, L.; Schmidt, M. and Fuhs, W. (2002), Interface recombination in amorphous/crystalline silicon solar cells, a simulation study, *Proceedings of 29th IEEE Photovoltaic Specialists' Conf.*,19-24 May (2002), New Orleans, USA, pp. 1238-1241.

Fuhs, W.; Niemann, K. and Stuke, J. (1994), Heterojunctions of amorphous silicon and silicon single crystals, *AIP Conference Proceedings*, Vol 20 (May 1974) pp 345-350.

Fujiwara, H.; Sai, H. and Kondo M. (2009), Crystalline Si Heterojunction Solar Cells with the Double Heterostructure of Hydrogenated Amorphous Silicon Oxide, *Japanese Journal of Applied Physics* Vol 48 (June 2009) pp 064506 -1-4, ISSN 0021-4922.

Guha, S.; Yang, J.; Pawlikiewicz, A.; Glatfelter, T., Ross, R. and Ovshinsky, S. R . (1989), Band-gap profiling for Improving the efficiency of amorphous silicon alloy solar cells Applied Physics Letters Vol 54, No 23 (June 1989) pp 2330-2332, ISSN 0021-8979.

Hack, M. and Schur, M (1985)., Physics of amorphous silicon alloy p-I-n solar cells, *Journal of Applied Physics*, Vol 58 (1985) pp 997-1020, ISSN 0021-8979.

Kanevce, A. and Metzger W. K. (2009), The role of amorphous silicon and tunneling in heterojunction with intrinsic thin layer (HIT) solar cells, *Journal of Applied Physics*, Vol 105, No. 9, (May,2009), pp. 094507-1-7, ISSN 0021-8979.

Leblanc, F.; Perrin, J. and Schmitt, J. (1994), Numerical modeling of the optical properties of hydrogenated amorphous-silicon-based p-i-n solar cells deposited on rough transparent conducting oxide substrate, *Journal of Applied Physics* Vol 75, No.2 (January 1994) pp 1074-1087, ISSN 0021-8979.

Maruyama, E.; Terakawa, A.; Taguchi, M.; Yoshimine, Y.; Ide, D.; Baba, T.; Shima, M.; Sakata, H. and Tanaka, M. (2006), *Sanyo's Challenges to the Development of High-efficiency HIT Solar Cells and the Expansion of HIT Business, Proceedings 4th World Conf. on Photovoltais Solar Energy Conversion*, Hawaii, USA, 9-12 May (2006).

McElheny, P.; Arch, J. K.; Lin, H.-S. and Fonash, S., Range of validity of the surface-photovoltage diffusion length measurement: A computer simulation *Journal of Applied Physics*, Vol 64 No. 3 (August 1988) pp 1254-1265, ISSN 0021-8979.

Meier, D.L.; Page, M.R.; Iwaniczko, E.; Xu, Y.; Wang, Q. and Branz, H.M. (2007), Determination of surface recombination velocities for thermal oxide and amorphous silicon on float zone silicon, *Proceedings of the 17th Workshop on Crystalline Silicon Solar Cells and Modules*, pp.214-217, Vail, CO, USA, 2007.

Nath, M; Chatterjee, P.; Damon-Lacoste, J. and Roca i Cabarrocas, P. (2008), Criteria for improved open-circuit voltage in a-Si:H(N)/c-Si(P) front heterojunction with intrinsic thin layer solar cells, *Journal of Applied Physsics* , Vol 103 (February, 2008) pp 034506-1-9, ISSN 0021-8979.

Olibet, S.; Vallat-Sauvain E.; Fesquet, L.; Monachon, C.; Hessler-Wyser, A.; Damon-Lacoste, J.; De Wolf, S. and Ballif, C. (2010), Properties of Interfaces in Amorphous/Crystalline Silicon Heterojunctions, *Physics Status Solidi A*, Vol 207, No. 3 (January, 2010)pp 651–656, ISSN 1862-6300.

Olibet, S.; Vallat-Sauvain, E. and Ballif, C. (2007), Model for a-Si:H/c-Si interface recombination based on the amphoteric nature of silicon dangling bonds, *Physical Review B*, Vol 76, No 3, (July, 2007), pp 035326-1 – 14, ISSN 1098-0121.

Osuda, K.; Okamoto, H. and Hamakawa, Y. (1983), Amorphous Si/Polycrystalline Si stacked solar cells having more than 12% conversion efficiency,*Japanese Journal of Applied Physics*, Vol 22, No.9 (September, 1983), pp. L605-L607, ISSN 0021-4922

Palit, N.; and Chatterjee, P. (1998), A computer analysis of double junction solar cells with a-Si : H absorber layers, *Solar Energy Materials & Solar Cells* , Vol 53 (1998), pp. 235-245, ISSN 0927-0248.

Plá, J.; Tamasi, M.; Rizzoli, R.; Losurdo, M.; Centurioni, E.; Summonte, C. and Rubinelli, F.(2003), Optimization of ITO layers for applications in a-Si/c-Si heterojunction solar cells , *Thin Solid Films*, Vol 425, No 1-2, (February 2003) pp. 185-192, ISSN 0040-6090.

Rahmouni, M.; Datta, A.; Chatterjee, P.; Damon-Lacoste, J.; Ballif, C. and Roca i Cabarrocas, P. (2010), Carrier transport and sensitivity issues in heterojunction with intrinsic thin layer solar cells on N-type crystalline silicon: A computer simulation study *Journal of Applied Physics*, Vol 107, No 5, (March, 2010) pp. 054521-1-14, ISSN 0021-8979.

Sakata, H.; Nakai, T.; Baba, T.; Taguchi, M.; Tsuge, S.; Uchihashi, K. and S. Kyama (2000), 20.7% highest efficiency large area (100.5 cm²) HITT^M cell, *Proceedings 28th IEEE Photovoltaic Spealist conference*, pp. 7-12, ISBN 0-7803-5772-8, Anchoorage, Alaska, 2000 September 15 -22.

Sawada, T.; Terada, N.; Tsuge, S.; Baba, T.; Takahama, T.; Wakisaka, K.; Tsuda, S. and Nakano S. (1994), High-efficiency a-Si/c-Si heterojunction solar cell, *Proceedings 1st World Conference on Photovoltaic Solar Energy Conversion*, pp.1219-1226, ISBN 0-7803-1460-3, Hawaii, USA, 1994, December 5-9.

Schmidt, M.; Korte, L.; Laades, A.; Stangl, R.; Schubert, Ch.; Angermann, H.; Conrad, E. and Maydell, K. V. (2007), Physical aspects of a-Si:H/c-Si hetero-junction solar cells , *Thin Solid Films*, Vol 515, No. 19 (July, 2007) pp. 7475-7480, ISSN 0040-6090.

Schmidt, M.; Angermann, H.; Conrad, E.; Korte, L.; Laades, A.; Maydell, K. V.; Schubert, Ch. and Strangl, R. (2006), Physical and Technological Aspects of a-Si:H/c-Si Hetero-Junction Solar Cells, *Proceedings of 4th World Conference on Photovoltaic Energy Conversion*, pp. 1433-1438, ISBN 1-4244-0017-1, Hawaii, USA, 2006, May, 7-12.

Sritharathikhun, J.; Yamamoto, H.; Miyajima, S.; Yamada, A. and Konagai, M. (2008), Optimization of Amorphous Silicon Oxide Buffer Layer for High-Efficiency p-Type Hydrogenated Microcrystalline Silicon Oxide/n-Type Crystalline Silicon Heterojunction Solar Cells, *Japanese Journal of Applied Physics*, Vol 47, No. 11, (November, 2008) pp. 8452-8455, ISSN 1347-4065.

Smole, F. and Furlan J. (1992), Effects of abrupt and graded a-Si:C:H/a-Si:H interface on internalproperties and external characteristics of p-i-n a-Si:H solar cells, *Journal of Applied Physics*, Vol 72, No. 12, (September, 1992) pp. 5964-5969, ISSN 0021-8979.

Stangl, R.; Froitzheim, A.; Elstner, L. and Fuhs, W. (2001), Amorphous/crystalline silicon heterojunction solar cells, a simulation study, *Proceedings of 17th European Photovoltaic Solar Energy Conference*, pp. 1387-1390, ISBN 3-936338-07-8, Munich, Germany, 2001, October 22-26.

Sze, S.M. (1981), *Physics of Semiconductor Devices* (2nd Edition),John Wiley & Sons, ISBN 9971-51-266-1.

Taguchi, M., Sakata, H.; Yoshihiro, Y.; Maruyama, E.; Terakawa, A.; Tanaka, M. and Kiyama, S. (2005), An approach for the higher efficiency in the HIT cells, *Proceedings of 31st IEEE Photovoltaic Specialists Conference*, pp.866 – 871, ISBN 0-7803-8707-4, Lake Buena Vista, FL, 2005, January 3-7.

Taguchi, M.; Maruyama, E. and Tanaka, M., Temperature Dependence of Amorphous/Crystalline Silicon Heterojunction Solar Cells, *Japanese Journal of Applied Physics* , Vol 47, (February, 2008), pp. 814-817, ISSN 0021-4922.

Takahama, T.; Taguchi, M.; Kuroda, S., Matsuyama, T.; Tanaka, M.; Tsuda, S.; Nakano, S. and Kuwano, Y. (1992), High efficiency single- and polycrystalline silicon solar cells using ACJ-HIT structure,*Procedings of 11th European Photovoltaic Solar Energy Conference*, pp. 1057-1062, Montreux, Switzerland, 1992, October 12-16.

Tanaka, M.; Taguchi, M.; Matsuyama, T.; Sawada, T.; Tsuda, S.; Nakano, S.; Hanafusa, H. and Kuwano, Y. (1992), Development of New a-Si/c-Si Heterojunction Solar Cells: ACJ-HIT (Artificially Constructed Junction-Heterojunction with Intrinsic Thin-Layer), *Japanese Journal of Applied Physics*, Vol 31, (November 1992), pp. 3518-3522, ISSN 0021-4922.

Tanaka, M.; Okamoto, S.; Tsuge, S. and Kiyama, S. (2003), Development of hit solar cells with more than 21% conversion efficiency and commercialization of highest performance hit modules, *Proceedings of 3rd World Conference on Photovoltaic Energy Conversion*, Vol 1, pp. 955–958, ISBN 4-9901816-0-3, Osaka. Japan, 2003, May 11-18.

Tchakarov S., Roca i Cabarrocas P., Dutta U., Chatterjee P. and Equer B., Experimental study and modeling of reverse-bias dark currents in PIN structures using amorphous and polymorphous silicon, *Journal of Applied Physics*, Vol 94, No. 11, (December, 2003) , pp. 7317-7327. ISSN (print) 0021-8979.

van Cleef, M. W. M.; Schropp, R. E. I. and Rubinelli, F. A. (1998a), Significance of tunneling in p^+ amorphous silicon carbide n crystalline silicon heterojunction solar cells, *Applied Physics Letters*, Vol 73, No 18, (November, 1998), pp. 2609-2611, ISSN 0003-6951.

van Cleef, M. W. M.; Rubinelli, F. A.; Rizzoli, R.; Pinghini, R.; Schropp, R. E. I. and. van der Weg, W. F. (1998b), Amorphous Silicon Carbide/Crystalline Silicon Heterojunction Solar Cells: A Comprehensive Study of the Photocarrier Collection, *Japanese Journal Applied Physics*, Vol 37, (July, 1998), pp. 3926-3932, ISSN 1347-4065.

Veschetti, Y.; Muller, J.-C.; Damon-Lacoste, J.; Roca i Cabarrocas, P.; Gudovskikh, A. S.; Kleider, J.-P.; Ribeyron, P.-J. and Rolland, E. (2006), Optimisation of amorphous and polymorphous thin silicon layers for the formation of the front-side of heterojunction solar cells on p-type crystalline silicon substrates, *Thin Solid Films*, Vol 511-512, (July, 2006), pp. 543-547, ISSN 0040-6090.

von der Linden M. B., Schropp R. E. I., van Sark W. G. J. H. M., Zeman M., Tao G. and Metselaar J. W., The influence of TCO texture on the spectral response of a-Si:H solar cells, *Proceedings of 11th European Photovoltaic Solar Energy Conference*, pp. 647 Montreux, Switzerland, 1992, October 12-16.

Wang, Q.; Page, M.; Yan, Y. and Wang, T. (2005), High-throughput approaches to optimization of crystal silicon surface passivation and heterojunction solar cells, *Proceedings of the 31st IEEE Photovoltaic Specialists Conference*, pp.1233-1236, ISBN 0-7803-8707-4, Orlando, FL, USA, 2005, January 3-7.

Wang, Q.; Page, M.R.; Iwaniczko, E.; Xu, Y.Q.; Roybal, L.; Bauer, R.; To, B.; Yuan, H.C.; Duda, A. and Yan, Y.F., Crystal silicon heterojunction solar cells by hot-wire CVD, *Proceedings of the 33rd IEEE Photovoltaic Specialists Conference*,pp 1-5, ISBN: 978-1-4244-1640-0, San Diego, CA, USA, 2008, May 11-16.

Wang, Q.; Page, M. R.; Iwaniczko, E.; Xu, Y.; Roybal, L.; Bauer, R.; To, B.; Yuan, H.-C.; Duda, A.; Hasoon, F.; Yan, Y. F.; Levi, D.; Meier, D.; Branz Howard, M. and Wang, T. H. (2010), Efficient heterojunction solar cells on p-type crystal silicon wafers, *Applied Physics Letters*, Vol 96, No. 1, (January, 2010), pp. 013507-1-3, ISSN 0003-6951.

Zeman, M.; van Swaaij, R. A. C. M. M.; Metselaar, J. W. and Schropp, R. E. I. (2000), Optical modeling of *a*-Si:H solar cells with rough interfaces: Effect of back contact and interface roughness, *Journal Applied Physics* , Vol 88, No. 11 (December, 2000) pp. 6436-6443, ISSN 0021-8979.

News release by SANYO on 22nd May, 2009, SANYO Develops HIT Solar Cells with World's Highest Energy Conversion Efficiency of 23.0%.

< http://panasonic.net/sanyo/news/2009/05/22-1.html>.

4

Fabrication of the Hydrogenated Amorphous Silicon Films Exhibiting High Stability Against Light Soaking

Satoshi Shimizu[1,2], Michio Kondo[1] and Akihisa Matsuda[3]
[1]Research Center for Photovoltaics, National Institute
of Advanced Industrial Science and Technology
[2]Max-Planck-Institut für extraterrestrische Physik
[3]Graduate School of Engineering Science,
Osaka University
[1,3]Japan
[2]Germany

1. Introduction

A hygrogenated amorphous silicon (a-Si:H) thin film solar cell was first reported in 1976 [Carlson, & Wronski, 1976]. Since then, intensive works have been carried out for the improvement of its performances. Attempt to increase the conversion efficiencies of the thin film solar cells, a multi junction solar cell structure was proposed and has been investigated [Yang et al., 1997; Shah et al., 1999; Green, 2003; Shah et al., 2004]. It consists of the intrinsic layers having different optical bandgaps in order to absorb the sunlight efficiently in a wide spectrum range.

The density of photo-generated carriers is determined by the light absorption coefficient and the defect density of a material. The absorption coefficient of a-Si:H in a visible light region is one order magnitude higher than that of μc-Si:H due to the direct transition phenomenon. Therefore, a thin a-Si:H layer absorbs sufficient photons. This is a huge advantage for the thin film based solar cell technology in which mass production should be definitely taken into account.

However, a-Si:H has another aspect known as a Staebler-Wronski effect, i.e., the number of unpaired Si dangling bonds increases with light soaking, which lowers photocarrier density by decreasing carrier lifetime [Staebler & Wronski, 1977]. Indeed, conversion efficiencies of a-Si:H based solar cells deteriorate generally by 15-20 % due to this phenomenon. On the other hand, it is possible to suppress this deterioration to some extent by reducing a film thickness of a-Si:H with efficient light-trapping structures [e.g., Müller et al., 2004]. Indeed, the fabrication of the highly stabilized a-Si:H single junction solar cell by the precise optimizations of the optical properties and the i-layer thickness has been reported [Borrello et al., 2011]. Besides those intensive efforts, establishing the technique for fabricating highly stable a-Si:H films is essentially very important to extract its maximum potential for the solar cell applications.

Phenomenologically, a good correlation is observed between degradation ratio of a-Si:H and its hydrogen concentration, namely Si-H$_2$ bond density where a low Si-H$_2$ bond density film exhibits high stability [Takai et el., 2000]. Although the detailed microscopic model for explaining this correlation has not been revealed yet, the tendency is observed in the films prepared under the wide range of fabrication conditions [Nishimoto et al., 2002]. One of the methods to reduce a hydrogen concentration is to increase a substrate temperature. However, a high processing temperature results in increasing initial defect density. Additionally, it is preferable to use the processing temperature of around or less than 200 ºC from the viewpoint of low cost fabrications. Reducing Si-H$_2$ bond density without increasing a substrate temperature is one of the key issues for the fabrication of stable a-Si:H films.

In a chemical vapor deposition process, there are mainly two steps to be considered, i.e., 1) gas phase reactions and 2) surface reactions. In the first step, depending on the electron temperature in a silane plasma, several types of precursors are generated, and they play an important role on the properties of resulting films [Matsuda, 2004]. For example, the a-Si:H films prepared under a powder rich gas condition have very high initial defect densities, namely at the low substrate temperatures [e.g., Roca i Cabarrocas, 2000]. Those powders or so-called higher-ordered silane radicals are created by the insertion reactions of SiH$_2$ radicals produced generally under a high electron temperature condition in a silane plasma. This insertion reaction is a rapid process. The SiH$_2$ radicals are created even under a relatively low electron temperature condition because it is statistically difficult to eliminate only high energy electrons from the system. A higher-ordered silane radical causes a steric hindrance and inhibits short range-ordered sp^3 bond formations on the film growing surface. For example, it is observed that the Si-H$_2$ bond density in the film, which has correlation with light-induced degradation of a-Si:H, increases when the density of the higher-ordered silane radicals in a gas phase is high [Takai et al., 2000].

In this work, to study the effect of precursors in a gas phase on the properties of the resulting film, a triode deposition system is applied for the growth of a-Si:H films where a mesh is installed between a cathode and a substrate. With such a configuration, a long lifetime radical such as SiH$_3$ mainly contributes to the film growth [Matsuda & Tanaka, 1986]. The properties and the stabilities of the resulting films are evaluated.

2. Fabrication and evaluation methods

The preparations of a-Si:H films were performed using a triode deposition system. Figure 1 shows the schematic of the system. A mesh is placed between the cathode and the substrate scepter in which a heater is mounted. VHF (100 MHz) voltage is applied on the cathode with the 20 sccm of SiH$_4$ gas flow, and a silane plasma is generated between the cathode and the negatively dc-biased mesh. All the films were prepared at 100 mTorr (13.3 Pa). The deposition precursors pass through the mesh and reach to the substrate. The substrate scepter is movable, and the distance between the mesh and the substrate (d_{ms}) is one of the important deposition parameters. The distance between the cathode and the mesh is fixed at 2 cm. In some cases, an additional mesh is installed behind the pre-existing mesh with the distance of 1.5 mm at which no plasma is generated between the two mesh under our conditions. The volume of the chamber is c.a. 1.1×10^4 cm^3, and its base pressure is c.a. 3×10^{-8} Torr. The diameters of the electrodes are 10 cm. As a comparison, a-Si:H films were also prepared with a conventional diode system where no mesh is installed. In this case, the distance between the cathode and the substrate is fixed at 2 cm.

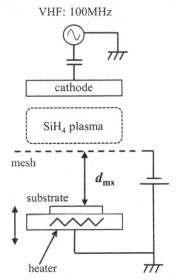

Fig. 1. Schematic of the a-Si:H growth chamber used in this study. A negatively dc-biased mesh is installed between the cathode and the substrate. The distance between the mesh and the substrate (d_{ms}) is adjustable.

The densities of Si–H and Si–H$_2$ bonds in the resulting film deposited on a intrinsic Si substrate were calculated from the integrated intensities of the stretching modes in a Fourier transform infrared spectroscopy (FTIR) spectrum, where the proportional constants are 9.0×10^{19} cm^2 for Si–H and 2.2×10^{20} cm^2 for Si–H$_2$, respectively [Langford et al., 1992]. The neutral spin density of the film deposited on a quartz substrate was measured by electron paramagnetic resonance (EPR). To study light-soaking stability of the film, a Schottky diode was fabricated on a phosphorous doped n$^+$Si substrate (0.03 Ωcm) with a half transparent Ni electrode on the top (n$^+$Si/a-Si:H/Ni). The native surface oxide layer on the n$^+$Si substrate was etched with diluted HF solution before the growth of a-Si:H.

A p-i-n structured solar cell (5×5 mm^2) was fabricated in a multi-chamber system. The doped layers were prepared in conventional diode system chambers, and the i-layer was fabricated in a triode system chamber at 180 °C. The distance between the mesh and the substrate is 1.5 cm. The other detailed conditions for the solar cell fabrication are described elsewhere [Sonobe et al., 2006]. The I–V characteristics of the solar cells were measured under an illumination of AM 1.5, 100 mW/cm^2 white light. In every case, the light degradation was performed by illuminating intense 300 mW/cm^2 white light for 6 h at 60 °C.

3. Properties and stabilities of the triode-deposited a-Si:H

3.1 Properties of the a-Si:H films prepared by the triode system
3.1.1 Hydrogen concentration
The hydrogen concentrations of the a-Si:H films prepared by the triode system were measured by FTIR. Figure 2 (a) shows the spectrum of the film prepared at 250 °C with the

distance between the mesh and the substrate, d_{ms}, of 3 cm [Shimizu et al., 2005]. As a comparison, that of the conventionally prepared a-Si:H film at the same substrate temperature is shown in figure 2 (b) [Shimizu et al., 2005]. One can see that the Si-H$_2$ bond density is low in the case of the triode deposition.

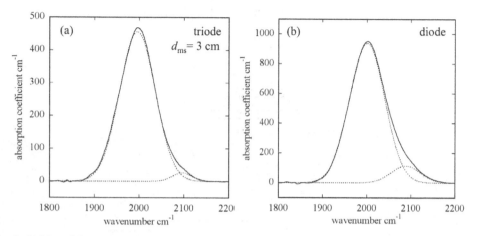

Fig. 2. Si-H and Si-H$_2$ stretching mode absorption spectra obtained in the FTIR measurement. The films were prepared by the: (a) triode system at the d_{ms} of 3 cm, and the (b) conventional diode system. In both cases, the substrate temperatures are 250 ºC. [Shimizu et al., 2005]

Furthermore, the a-Si:H films were fabricated with changing d_{ms}, and the results are summarized in figure 3. One can see that, as d_{ms} is increased, both Si-H and Si-H$_2$ densities decrease. The a-Si:H film prepared at d_{ms} = 4 cm contains the Si-H bond density of 4.0 at.% and less than 0.1 at.% (close to the detection limit of FTIR) of Si-H$_2$ bond density. On the other hand, the film prepared by the conventional diode method contains 9.0 at.% of Si-H bonds and 1.5 at.% of Si-H$_2$ bonds at the same substrate temperature. The similar reductions of Si-H and Si-H$_2$ bond densities with the triode system are observed in the films prepared under the several substrate temperatures as shown in figure 3.

3.1.2 Growth of a-Si:H with double mesh

With installing a mesh and increasing d_{ms}, the growth rate is reduced. To see the effect of growth rate on the resulting hydrogen concentration, the films were prepared with installing a second mesh at a fixed VHF input power and d_{ms}. With such a configuration, one can control the growth rate without changing the gas phase conditions, whereas it is not the case if the VHF power or d_{ms} is changed to control the growth rate, because the generation rate of precursors changes with the input power, and as discussed later, d_{ms} affects the flux of the precursors reaching to the substrate. Thus, to see the effect of the growth rate, the double mesh configuration was used.

Here, the films were prepared with or without the second mesh, which is represented as double or a single mesh, respectively. All the films were prepared at 250 ºC. The results are summarized in table 1 and figure 4 [Shimizu et al., 2007]. At the VHF power of 10 W, almost the same hydrogen concentrations are observed both in the single and the double mesh

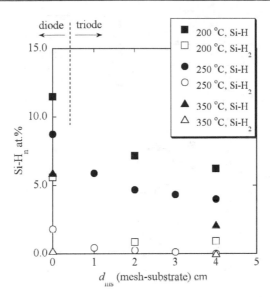

Fig. 3. Si-H and Si-H$_2$ bond densities in the a-Si:H films fabricated with the triode deposition system (triode) under the various distances between the mesh and the substrate (d_{ms}). As a comparison, those of the conventionally prepared films without the mesh are also shown (diode, d_{ms} = 0 cm). The films were prepared at the substrate temperatures of 200, 250 and 350 ºC, respectively.

cases, but the growth rates are different each other where very low growth rate is observed with the double mesh. The growth rate with the double mesh at 10 W is c.a. 0.1 Å/s which is close to the value observed at the VHF power of 2 W with the single mesh. However, the observed Si-H and Si-H$_2$ bond densities are lower in the case of 2 W with the single mesh. The similar trend is observed under the different conditions as shown in figure 4.

input power (W)	mesh	growth rate (Å/s)	Si-H (at.%)	Si-H$_2$ (at.%)
2	single	0.18	4.0	< 0.1
10	double	0.12	6.1	0.9
10	single	0.80	6.6	1.0

Table 1. Si-H and Si-H$_2$ bond densities and the observed growth rate of the films prepared under the several conditions with fixing d_{ms} (= 4 cm) and the substrate temperature (= 250 ºC).

3.1.3 Microscopic structure
In figures 5 (a) and (b), the FWHM of the Si-H and Si-H$_2$ stretching mode peaks in the FTIR spectra are platted against the density of Si-H and Si-H$_2$, respectively. The films were prepared at the VHF input power of 2 or 10 W using the each electrode configuration i.e., triode or diode system as indicated in the figure. The substrate temperature is 250 ºC in every case. While the scattered relation is observed in the Si-H bond case, one can see the good correlation between the Si-H$_2$ bond densities and their FWHMs. Moreover, while

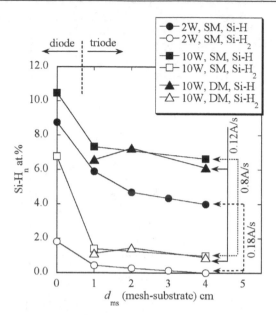

Fig. 4. Si-H and Si-H$_2$ bond densities in the a-Si:H films fabricated under the various conditions. Open and closed circle: VHF = 2 W, with a single mesh (2 W, SM), open and closed square: VHF = 10 W, with a single mesh (10 W, SM), open and closed triangle: VHF = 10 W, with double mesh (10 W, DM). As a comparison, those of the conventionally prepared films without the mesh are also shown (diode, d_{ms} = 0 cm). [Shimizu et al., 2007]

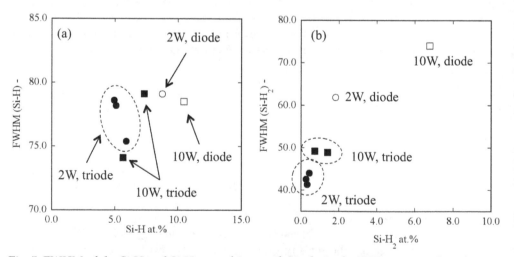

Fig. 5. FWHM of the Si-H and Si-H$_2$ stretching mode peaks in the FTIR spectra platted against the density of Si-H and Si-H$_2$, respectively. The films were prepared at the VHF input power of 2 or 10 W using the each electrode configuration (triode or diode) as indicated.

the FWHM values of the S-H peaks are more or less in the same range, the narrower FWHMs of the Si-H$_2$ peaks are observed in the triode-deposited films. Furthermore, when the electrode configuration is the same (triode or diode), the films prepared at the lower VHF input power exhibit narrower FWHMs of the Si-H$_2$ peaks.

3.1.4 Conductivity

The conductivities of the a-Si:H films fabricated using the triode system are measured. Figure 6 shows the dark and photoconductivities of the films. The photoconductivity was measured under the illumination of 100 mW/cm^2 white light. The observed dark-conductivities are of the order of 10^{-11} S/cm. The deposition rate of the triode system is typically less than 1 Å/s, which may cause unfavorable impurity incorporations during the film growth, causing the reduction of photosensitivity due to the increase of dark-conductivity. The dark-conductivity of the triode-deposited a-Si:H is, however, in the range equivalent to that observed in the diode-deposited film grown at 7.3 Å/s, and the photoconductivities of those films are of the order of 10^{-5} S/cm. The result indicates that the triode-deposited a-Si:H films do not contain substantial number of impurities which deteriorates photosensitivity.

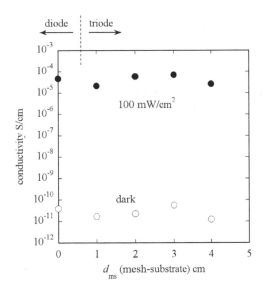

Fig. 6. The dark and photoconductivities of the a-Si:H films prepared either by a triode or a diode deposition system (d_{ms} = 0 cm).

3.2 Stabilities of the triode-deposited a-Si:H films
3.2.1 Spin density

Degradation of the film prepare by the triode system is checked by measuring the change of neutral spin density by light soaking. Figure 7 shows the result [Shimizu et al., 2008]. All the films were prepared at 250 ºC, and as a comparison, the results of the diode-deposited films are also shown. The spin density is plotted against Si-H$_2$ bond density. The initial defect densities are almost the same throughout the samples ($\approx$ 2×10^{15} cm^{-3}). On the other hand, more stable behaviors are observed in the triode-deposited a-Si:H films in the degraded states. The trend is best seen in the film prepared at the d_{ms} of 4 cm where the lowest Si-H$_2$ bond density is observed as shown in figure 3.

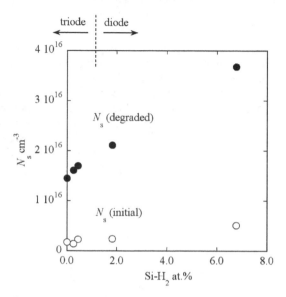

Fig. 7. Change in the neutral spin density (N_s) due to light soaking as a function of Si-H$_2$ bond density in the film. [Shimizu et al., 2008]

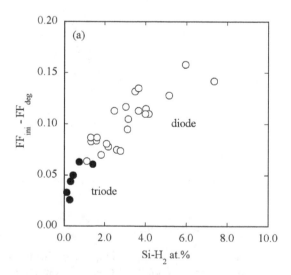

Fig. 8. Light-induced change in the fill-factor ($\Delta FF = FF_{ini} - FF_{deg}$) of the Schottky diode having the intrinsic layer produced at the each condition. Closed circle: triode-deposited film (triode), open circle: conventionally prepared film (diode). [Shimizu et al., 2005]

3.2.2 Schottky diode

Furthermore, the stabilities of the triode-deposited a-Si:H films were studied with fabricating the Schottky diodes where their fill-factor (FF) changes were evaluated as a measure of degradation. The intrinsic layer of the Schottky diode was fabricated either by a triode or a diode system under the various conditions. The fill-factors in the initial state (FF_{ini}) are almost the same throughout the samples: 52 - 54 %. On the other hand, the fill-factors in the degraded state (FF_{deg}) are different each other. In figure 8, the change in the fill-factor ($\Delta FF = FF_{ini} - FF_{deg}$) is plotted against Si-$H_2$ bond density [Shimizu et al., 2005]. For comparison, those of the films prepared with the diode system under the various conditions are also shown [Nishimoto et el., 2002]. One can see that the triode-deposited a-Si:H films contain low Si-H_2 bond densities, and correspondingly, the observed ΔFFs are low. Note that, the scattered correlation is observed when ΔFFs are plotted against the Si-H densities of the films [Shimizu et al., 2005].

3.2.3 Solar cell

The stability of the triode-deposited a-Si:H is checked with fabricating a p-i-n solar cell where the i-layer is deposited with a triode system. Since a multi chamber was used to prepare the solar cell, the i-layer fabrication conditions including the chamber geometry are different from those used in the previous sections. Especially, the distance between the mesh and the substrate is short as 1.5 cm which lowers the effect of Si-H_2 bond elimination than that achieved at larger distances as shown in figure 3. Additionally, the i-layer growth temperature of 180 ℃ was chosen. Therefore, the Si-H_2 bond density in the i-layer is slightly high as indicated in figure 3. On the other hand, we chose this temperature from the viewpoint of the device applications in which low temperature operations are preferable. The i-layer thickness is 250 nm. The I-V characteristic of the solar cell is shown in figure 9 [Sonobe et al., 2006].

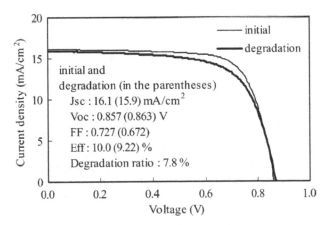

Fig. 9. The I-V characteristic of the p-i-n solar cell. The i-layer was prepared with the triode system at the substrate temperature of 180 ℃. The distance between the mesh and the substrate is 1.5 cm. [Sonobe et al., 2006]

The initial conversion efficiency is 10.0 %, and after the light soaking, the stabilized efficiency of 9.2 % is achieved. The degradation ratio is 7.8 % which is the lower value compared with that generally observed in the a-Si:H solar cell prepared by a conventional method with the same i-layer thickness. While further optimization is necessary to achieve higher stabilized efficiency, the result demonstrates the low degradation ratio of the a-Si:H solar cell with improving the stability of the i-layer itself, which is one of the essential solutions to obtain a stable a-Si:H solar cell.

4. Hydrogen elimination process

4.1 Hydrogen elimination process – post annealing

It is observed that the films grown by the triode system contain very low hydrogen concentrations, namely $Si-H_2$ bond densities. Those values change with the distance between the mesh and the substrate where the lowest hydrogen concentration is observed at the largest distance between the mesh and the substrate. In this section, we will discuss the possible mechanism for the reduction of Si-H and $Si-H_2$ bond densities in the triode deposition system.

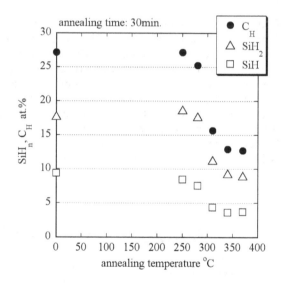

Fig. 10. Thermal effusion of hydrogen from the a-Si:H films deposited at 110 °C. The C_H is the sum of the Si-H and the $Si-H_2$ bond densities [Shimizu et al., 2007].

Hydrogen elimination takes place both in a film growth state and in a post annealing state when a substrate temperature is high. To distinguish it in our case, at first, the thermal annealing tests were performed on the a-Si:H films prepared at the low substrate temperature of 110 °C using the diode system. The as-deposited films contain the large initial hydrogen concentrations (C_H) of c.a. 27 at.%. After the growth, the individual film was kept in the deposition chamber and was annealed for 30 minutes at the certain temperature. The result is shown in figure 10 [Shimizu et al., 2007]. One can see that the hydrogen concentration is reduced at the high annealing temperatures. On the other hand, at the temperature of 250 °C, which is the substrate temperature used in our triode deposition system, no C_H reduction takes place at least from the bulk. The result shows that under the substrate temperature of 250 °C, the hydrogen elimination process takes place during the film growth, i.e., most likely with gas reactions.

4.2 Hydrogen elimination process during film growth

The possible hydrogen elimination processes during the a-Si:H film growth are the following and are schematically shown in figure 11.

a. hydrogen abstraction reaction by an atomic hydrogen
b. spontaneous thermal desorption of surface hydrogen
c. hydrogen abstraction reaction by a SiH$_3$ radical
d. hydrogen elimination process through a cross-linking reaction

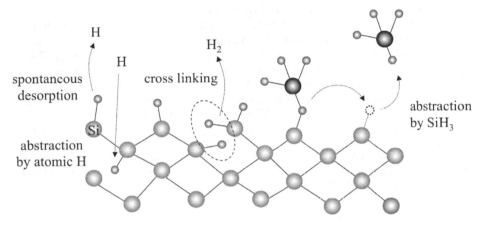

Fig. 11. Schematic of the hydrogen elimination processes during the growth of a-Si:H.

a. Hydrogen abstraction reaction by an atomic hydrogen

Atomic hydrogen exists in an silane plasma [e.g., Matsuda, 2004]. It reacts with a bonded hydrogen of a film and forms H$_2$ molecule, resulting in a hydrogen elimination. The probability of this reaction should be proportional to the flux of atomic hydrogen. In a silane plasma, generated radicals and ions collide with SiH$_4$ molecule of which density is high in the gas phase. When the atomic hydrogen reacts with SiH$_4$, SiH$_3$ radical and H$_2$ molecule are generated at the rate constant of ~ 3×10^{-12} cm^3/s [Kushner, 1988; Perrin et al., 1996]:

$$H + SiH_4 \rightarrow SiH_3 + H_2. \tag{1}$$

The stable H$_2$ molecule does not contribute to the abstraction of the bonded hydrogen. In the triode system, basically no atomic hydrogen is generated but only the collisions take place in the region between the mesh and the substrate, indicating that the density of atomic hydrogen near the substrate is low. Therefore, it is natural to say that the hydrogen elimination process is not dominated by atomic hydrogen in the triode system.

b. Spontaneous thermal desorption of surface hydrogen

The hydrogen desorption process from Si-H bond has been studied elsewhere [Toyoshima et al., 1991]. The activation energy of this reaction is estimated as 2 - 3 eV, and the reaction takes place only in the temperature range higher than 400 °C [Beyer & Wagner, 1983]. Therefore, it is unlikely that the spontaneous hydrogen desorption takes place under the substrate temperature of 250 °C as in our case.

c.   Hydrogen abstraction reaction by a SiH_3 radical

It is reported that the dominant deposition precursor for a-Si:H growth is a SiH_3 radical [Matsuda, 2004]. When a SiH_3 radical reaches to a growing surface, it physisorbs to one of the surface hydrogen atoms under a certain probability [Perrin et al., 1989; Matsuda et al., 1990].

$$\equiv SiH + SiH_3 \rightarrow \ \equiv SiHSiH_3. \tag{2}$$

The physisorbed SiH_3 radical diffuses on the surface with changing the physisorption spot, and it finally captures one of the surface hydrogen, forming SiH_4 and it leaves from the surface. As a result, hydrogen is abstracted and a surface Si dangling bond is created. When another surface diffusing SiH_3 radical reaches to the surface dangling bond, it is chemisorbed and a Si-Si bond is formed [Perrin et al., 1989; Matsuda et al., 1990]. Note that, if this dangling bond is not terminated with some radicals, defect density is increased in the resulting film. As one can see, as long as a SiH_3 radical is supplied to the dangling bond site after the hydrogen elimination by another SiH_3, the density of the surface hydrogen does not decrease through the process since the chemisorped SiH_3 also contains hydrogen.

d.   Hydrogen elimination process through a cross-linking reaction

A cross-linking hydrogen elimination reaction takes place with a pair of Si-H bonds facing each other. Hydrogen is eliminated with forming a H_2 molecule and a Si-Si bond [Matsuda & Tanaka 1986; Perrin et al., 1989],

$$\equiv Si-H + H-Si \equiv \rightarrow \ \equiv Si-Si \equiv + H_2. \tag{3}$$

Different from the other cases (a)–(c), a dangling bond does not remain after this reaction, therefore, an atomic hydrogen or a SiH_3 radical which contains hydrogen atom(s) does not stick to this site, resulting in a reduction of hydrogen concentration. As described above, a Si-Si bond is formed through this process, therefore, before the reaction, the both Si atoms should be located at the configuration of five or six members of ring to form a stable Si-Si bond. On the other hand, if two Si atoms are located apart from each other where the free energy of the resulting Si-Si bond including its surrounding structures after the reaction is higher than that of before, it is unlikely that the cross-linking reaction takes place. Such an inhabitation can occur when a higher-ordered silane radical sticks to a growing surface as discussed in the next section.

4.3 Effect of deposition precursors on hydrogen elimination

For the growth of a-Si:H film, a SiH_3 radical is a dominant species. On the other hand, higher-ordered silane radicals are also generated in a plasma through insertion reactions of SiH_2 [Takai et al., 2000]. It has been reported that when the higher-ordered silane radicals such as Si_4H_9 are incorporated into the film, Si-H_2 bond density increases [Takai et al., 2000; Nishimoto et al., 2002]. Therefore, the flux of those species toward the substrate determines the property of the resulting film. The diffusion length (L) of a species is

$$L = (D\tau)^{1/2}, \tag{4}$$

$$D \propto 1/m. \tag{5}$$

where D is a diffusion coefficient, t is a lifetime and m is a mass of a species.

First of all, the diffusion coefficient of a SiH$_3$ radical is larger than that of a higher-ordered silane radical due to the difference of their mass. The ratio of the mass of a Si$_n$H$_{2n+1}$ radical ($m_{SinH2n+1}$) and that of a SiH$_3$ radical (m_{SiH3}) is

$$m_{SinH2n+1} / m_{SiH3} \approx n. \tag{6}$$

Therefore, the diffusion coefficient of a SiH$_3$ radical is n-times larger than that of a Si$_n$H$_{2n+1}$ radical.

Second, the lifetime of a SiH$_3$ radical is especially long in a SiH$_4$ gas phase. The density of SiH$_4$ molecule under the condition used in this study (0.1 Torr, 250 ^{o}C) is $\sim 10^{15}$ cm^{-3}. The density of SiH$_3$ radicals is the highest among the other generated species, and the value is $\sim 10^{12}$ cm^{-3} in an RF silane plasma [Matsuda, 2004]. In the case of VHF plasma, as in our case, the value changes due to a high electron density and a low electron temperature effects. Estimating from the deposition rate, SiH$_3$ density can be one order of magnitude higher than that in an RF plasma, but in any cases, the density is still very low with respect to that of SiH$_4$ molecule. Therefore, most of the generated species collide with SiH$_4$. Here, the SiH$_3$ radical does not disappear due to the collision, resulting in its long lifetime:

$$SiH_3 + SiH_4 \rightarrow SiH_4 + SiH_3. \tag{7}$$

After all, the diffusion length of SiH$_3$ radical is very long according to equations (4) and (5). Under a certain gas flow rate condition, the radicals having small diffusion lengths are pumped out, but a large diffusion length species can still reach to the substrate. Thus, larger the distance between the mesh and the substrate, stronger the diffusion length effect. Therefore, we would like to propose that in the triode configuration, a long lifetime SiH$_3$ radical mainly contribute to the film growth than that in a diode system. If a SiH$_3$ radical sticks to a film surface, a cross-linking reaction takes place because the configuration of a five or six members of ring is easily formed with the SiH$_3$. On the other hand, when a higher-ordered silane radical sticks to the surface, not all of the Si atoms are located in such a configuration due to its steric-hindrance. Thus, the cross-linking reaction can take place partially, resulting in the remaining of hydrogen in the film. Once hydrogen is incorporated into the bulk of a film, it cannot be thermally eliminated at 250 ^{o}C as shown in figure 10. Indeed, under the low SiH$_4$ flow rate condition, the higher Si-H and Si-H$_2$ bond densities are observed in the triode system [Shimizu et al., 2007].

4.4 Effect of growth rate

In the case of the triode system, the growth rate is low compared to that observed in a conventional diode system. In our VHF plasma case, the growth rate observed with the diode system is 7.3 Å/s, and that observed in the triode system is 0.7 Å/s at d_{ms} = 1 cm, and is 0.2 Å/s at d_{ms} = 4 cm. When the growth rate is low, hydrogen concentration of the resulting film can be low when thermal desorption from the surface and the bulk are the dominant hydrogen elimination processes. However, such elimination processes at the substrate temperature of 250 ^{o}C are unlikely as discussed in the previous sections.

To confirm the effect of the growth rate furthermore, the experiments were performed with installing a second mesh as described in section 3.1.2. Installing the second mesh reduces the growth rate drastically. As shown in Table 1, the observed growth rate with the double mesh at the VHF power of 10 W is c.a. 0.1 Å/s and that with the single mesh at the same VHF power is 0.8 Å/s. Under those conditions, however, almost the same Si-H and Si-H$_2$ bond densities are observed. Since the VHF power is fixed at the same value, the densities of the generated radicals and ions in the plasma are basically the same in the both cases. When the VHF power is reduced to 2 W with a single mesh, the observed growth rate is 0.2 Å/s which is the similar value observed at the VHF power of 10 W with the double mesh (0.1 Å/s). On the other hand, the observed Si-H and Si-H$_2$ bond densities are different each other where lower values are observed in the low power case in which less higher-ordered silane radicals are produced due to a low electron temperature effect [Matsuda, 2004]. The results indicate that the gas phase condition is very important for determining a hydrogen concentration in the resulting film. Note that, when a plasma is unstable even in the triode case due to the lack of electrical matching, the observed Si-H and Si-H$_2$ densities are higher than the expected values shown in figure 3 (the higher value date are not presented).

5. Prospects for the future applications

The properties and the stabilities of the a-Si:H films prepared by the triode deposition system have been demonstrated in this study. The quality of the film is very good, and it exhibits very high stability against light soaking. Although the triode method reduces a growth rate of a film due to its configuration, which is a disadvantage for mass productions, we used this system to study the fundamental features. Several results indicate that the control of the gas phase condition is one of the essential factors to obtain a stable a-Si:H film. Based on this knowledge, one could establish the alternative fabrication methods which can produce the preferable gas phase condition for a stable a-Si:H fabrication.

In our result, the degree of degradation correlates well with Si-H$_2$ bond density in the film, which also corresponds to the former works. Beside the possibility of micro-void structure formation, one can propose the existence of chain-like Si-Si structures when the film contains large Si-H$_2$ bond density. Since it is a flexible structure, it can cause instability against light soaking. In figure 12 the ΔFF (= FF_{ini} − FF_{deg}) of the Schottky diode is plotted against the Si-H$_2$ bond density. It is a re-plot of figure 8 in a semi-log scale. The extrapolated line shows that the ΔFF value is zero at the Si-H$_2$ bond density of c.a. 1.3×10^{19} cm^{-3} (≈ 0.03 at.%). Although it is a hypothesis, the correlation indicates the guideline for the fabrication of stable a-Si:H films.

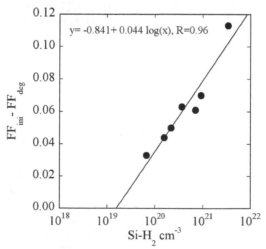

Fig. 12. Light-induced change in the fill-factor ($\Delta FF = FF_{ini}-FF_{deg}$) of the Schottky diode as a function of the Si-H$_2$ bond density (re-plot of fig. 8 in a semi-log scale).

6. Conclusions

Stable a-Si:H films against light soaking are prepared with adopting a triode deposition method where a mesh is placed between a cathode and a substrate. The resulting films contain very low Si-H and Si-H$_2$ bond densities compared with those observed in the films prepared by a conventional diode electrode method at the same substrate temperature. The hydrogen reduction effect is higher when the distance between the mesh and the substrate is increased. The films exhibit low initial defect densities and high photosensitivities. After the light soaking, high stabilities are observed in the films prepared by the triode system. The high stabilities of the films are also confirmed with the device configurations. It is most likely that the density of the precursors that reach to the growing surface is different each other in the triode and the diode systems. Control of gas phase condition is one of the key issues to fabricate stable a-Si:H films and related solar cells.

7. Acknowledgement

The authors acknowledge research support from the New Energy and Industrial Technology Development Organization (NEDO), Japan.

8. References

Beyer, W. & Wagner, H. (1983). The role of hydrogen in a-Si:H - Results of evolution and annealing studies. J. Non-Cryst Solids, Vol. 59-60, pp. 161-168.

Borrello, D., Vallat-Sauvain, E., Bailat, J., Kroll, U., Meier, J., Benagli, S., Marmelo, M., Monteduro, G., Hoetzel, J., Steinhauser & J., Lucie, C. (2011). High-efficiency amorphous silicon photovoltaic devices. WIPO Patent: WO/2011/033072.

Carlson, D. E. & Wronski, C. R. (1976). Amorphous silicon solar cell. Appl. Phys. Lett. Vol. 28, pp. 671-673.

Drevillon, B. & Toulemonde, M. (1985). Hydrogen content of amorphous silicon films deposited in a multipole plasma. J. Appl. Phys., Vol. 58, pp. 535-540.

Green, M. A. (2003). Crystalline and thin-film silicon solar cells: state of the art and future potential. Solar Energy, Vol. 74, pp. 181-192.

Kushner, M. J. (1988). A model for the discharge kinetics and plasma chemistry during plasma enhanced chemical vapor deposition of amorphous silicon. J. Appl. Phys., Vol. 63, pp 2532-2551.

Langford, A. A., Fleet, M. L., Nelson, B. P., Lanford, W. A. & Maley, N. (1992). Infrared absorption strength and hydrogen content of hydrogenated amorphous silicon. Phys. Rev. B, Vol. 45, pp. 13367-13377.

Matsuda, A. & Tanaka, K. (1986). Investigation of the growth kinetics of glow-discharge hydrogenated amorphous silicon using a radical separation technique. J. Appl. Phys., Vol. 60, pp. 2351-2356.

Matsuda, A., Nomoto, K., Takeuchi, Y., Suzuki, A., Yuuki, A. & Perrin, J. (1990). Temperature dependence of the sticking and loss probabilities of silyl radicals on hydrogenated amorphous silicon. Surf. Sci., Vol. 227, pp. 50-56.

Matsuda, A. (2004). Microcrystalline silicon. Growth and device application. J. Non-Cryst. Solids. Vol. 338–340, pp. 1–12.

Müller, J., Rech B., Springer, J. & Vanecek, M. (2004). TCO and light trapping in silicon thin film solar cells. Solar Energy, Vol. 77, pp. 917–930.

Nishimoto, T., Takai, M., Miyahara, H., Kondo, M. & Matsuda, A. (2002). Amorphous silicon solar cells deposited at high growth rate. J. Non-Cryst. Solids, Vol. 299-302, pp. 1116-1122.

Perrin, J., Takeda, Y., Hirano, N., Takeuchi, Y. & Matsuda, A. (1989). Sticking and recombination of the SiH₃ radical on hydrogenated amorphous silicon: The catalytic effect of diborane. Surf. Sci., Vol. 210, pp. 114-128.

Perrin, J., Leroy, O. & Bordage, M. C. (1996). Cross-sections, rate constants and transport coefficients in silane plasma chemistry. Contrib. Plasma Phys., Vol. 36, pp. 3-49.

Roca i Cabarrocas., P. (2000). Plasma enhanced chemical vapor deposition of amorphous, polymorphous and microcrystalline silicon films. J. Non-Cryst. Solids, Vol. 266-269, pp. 31-37.

Shah, A., Torres, P., Tscharner, R., Wyrsch, N. & Keppner, H. (1999). Photovoltaic Technology: The Case for Thin-Film Solar Cells., Science, Vol. 285, pp. 692-698.

Shah, A. V., Schade, H., Vanecek, M., Meier, J., Vallat-Sauvain, E., Wyrsch, N., Kroll, U., Droz, C. & Bailat, J. (2004). Thin-film Silicon Solar Cell Technology. Prog. Photovolt: Res. Appl., Vol. 12, pp. 113–142.

Shimizu, S., Kondo, M. & Matsuda, A. (2005). A highly stabilized hydrogenated amorphous silicon film having very low hydrogen concentration and an improved Si bond network. J. Appl. Phys., Vol. 97, pp. 033522 1-4.

Shimizu, S., Matsuda, A. & Kondo, M. (2007). The determinants of hydrogen concentrations in hydrogenated amorphous silicon films prepared using a triode deposition system. J. Appl. Phys., Vol. 101, pp. 064911 1-5.

Shimizu, S., Matsuda, A. & Kondo, M. (2008). Stability of thin film solar cells having less-hydrogenated amorphous silicon i-layers. Solar Energy Mater. & Solar Cells, Vol. 92, pp. 1241-1244.

Sonobe, H., Sato, A., Shimizu, S., Matsui, T., Kondo, M. & Matsuda, A. (2006). Highly stabilized hydrogenated amorphous silicon solar cells fabricated by triode-plasma CVD. Thin Solid Films, Vol. 502, pp. 306–310.

Staebler, D. L. & Wronski, C. R. (1977). Reversible conductivity changes in discharge-produced amorphous Si. Appl. Phys. Lett., Vol. 31, pp. 292-294.

Takai, M., Nishimoto, T., Takagi, T., Kondo, M. & Matsuda, A. (2000). Guiding principles for obtaining stabilized amorphous silicon at larger growth rates. J. Non-Cryst. Solids., Vol. 266-269, pp. 90-94.

Toyoshima, Y., Arai, K., Matsuda, A. & Tanaka, K. (1991). In situ characterization of the growing a-Si:H surface by IR spectroscopy. J. Non-Cryst. Solids, Vol. 137-138, pp. 765-770.

Yang, J., Banerjee, A. & Guha, S. (1997). Triple-junction amorphous silicon alloy solar cell with 14.6% initial and 13.0% stable conversion efficiencies. Appl. Phys. Lett., Vol. 70, pp. 2975-2977.

5

Large Area a-Si/μc-Si Thin Film Solar Cells

Fan Yang
Qualcomm MEMS Technologies, Inc.
United States

1. Introduction

Providing a sustainable and environment friendly energy source, photovoltaic (PV) power is becoming ever-increasingly important, as it decreases the nation's reliance on fossil-fuel generated electricity. Though widely regarded as a clean and renewable energy source, large scale deployment of PV is still impeded by the fact that the cost of PV energy is generally higher compared to grid electricity. Current development of PV technology is focused on two aspects: 1) improving the efficiency of PV modules and systems and 2) lowering the cost of delivered electricity through decreasing the manufacturing and installation cost. The merit of commercial solar cells aiming at terrestrial application is justified by the cost of unit PV power generation, dollar per watt ($/Wp), where Wp stands for the peak power generated by the cells.

Since the first practical PV cell grown on Si wafer at the Bell Laboratory in 1954, PV technology has been developed for more than five decades and evolved three "generations" based on different PV materials. The first generation of solar cells use crystalline materials, where the cost of the bulk materials has hit the point that further cost reduction is very difficult (Green 2007). In contrast, the second generation cells use thin film materials, where the required amount of materials is merely a few percent of that of bulk materials, significantly reducing the fabrication cost of this type of cells. The emerging, third generation of PV technology applies new materials and novel device concepts aiming at even higher efficiency and lower cost. At this moment, the commercial PV market is dominated by the first and second generation PV modules, and the third generation cells are still under lab research. As shown in Fig. 1, the efficiency of thin film PV system has improved from ~4 % in 1995 to >11 % in 2010, and will keep increasing to ~12% by 2020, a three-fold improvement compared to the system efficiency back in 1995. During the same timeframe, the cost of thin film PV system drops from ~4 $/Wp to ~0.5 $/Wp. Crystalline PV systems, though with higher efficiencies, have higher cost, i.e. 2.5 times the cost of thin film PV system. From the cost and material supply point of view, thin film solar cells will have a long-term development and gradually take more market share from the crystalline cells.

Many thin film materials can be used for PV cells, e.g., Si, CdTe, CIGS or the emerging organic/polymeric materials. Comparing to other materials, thin film Si, including amorphous Si (a-Si) and microcrystalline Si (μc-Si), have the following characters:

1. The PV active Si is the most abundant solid state element on the earth's shell, allowing for practically unlimited production of Si cells.

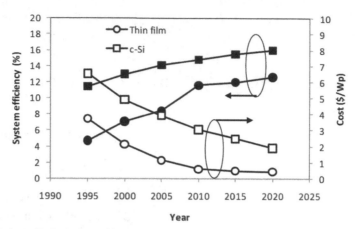

Fig. 1. Photovoltaic (PV) system efficiency and cost. Data from the U.S. Department of Energy.

2. Si has no toxicity and is environmental friendly.
3. Process of a-Si/μc-Si thin films takes the advantage of the highly mature semiconductor and display industries.
4. a-Si is a metastable material, and the initial cell performance of a-Si based cells degrades under illumination and then stabilizes, known as the Staebler–Wronski effect (Kolodziej 2004).

In addition, production of a-Si/μc-Si solar panels has a low entry barrier, thus making it more acceptable for the emerging PV manufactures. The first thin film Si solar cells were put into production in the 1980's when they were used as power sources for small electronic gadgets. Volume production of a-Si based solar panels started after the year 2000 with the introduction of large-area chemical vapor deposition (CVD) process at these companies: Sharp Corporation, United Solar Ovonic, Kaneka, Mitsubishi Heavy Industries, Ltd, etc. The true burst of Si thin film solar cells, on the other hand, came after 2007 with the "turnkey" (ready to use) thin-film solar manufacturing equipments introduced by Unaxis SPTec (later Oerlikon Solar) (Meier et al. 2007) and Applied Films Gmbh & Co. (later part of Applied Materials Inc.) (Repmann et al. 2007). The idea is that instead of developing the film deposition and module manufacturing technologies by self, the would-be solar maker can buy the full set of equipments together with the process recipes, and start manufacturing panels with relative ease. Each having a designed capacity of 40 – 60 MW, over twenty "turnkey" systems were sold to solar module makers world wide by Oerlikon and Applied Materials by 2010. The fast expansion of production capacity directly induced the drop of a-Si/μc-Si panel cost from around 5 $/Wp to less than 2 $/Wp.

At the moment thin film Si cells, including a-Si and μc-Si, take the largest market share (more than half of total production volume) among all types of thin film cells. Close to 5 GW of a-Si/μc-Si panels were manufactured in 2010, and will keep similarly large market share to at least 2013 (Fig. 2) (Young 2010). It is also noted from the same figure that the production volume of a-Si panels has an impressive compound annual growth rate (CAGR) of 42%, highest among all thin film PV technologies. Currently a significant amount of Si thin film panels are single-junction a-Si panels, whose efficiency will gradually increase to 8% - 8.5%. By adopting the a-Si/μc-Si multi-junction cells, panel efficiency will move up to

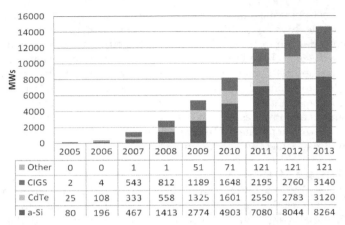

	2005	2006	2007	2008	2009	2010	2011	2012	2013
▪ Other	0	0	1	1	51	71	121	121	121
▪ CIGS	2	4	543	812	1189	1648	2195	2760	3140
▪ CdTe	25	108	333	558	1325	1601	2550	2783	3120
▪ a-Si	80	196	467	1413	2774	4903	7080	8044	8264

Fig. 2. Global thin film solar panel manufacturing capacity and compound annual growth rate (CAGR) by technology, 2006-2013 (estimate) (Young 2010).

around 10% after 2012. Costs for these technologies are expected to range from 0.80 to 1.20 $/Wp (Mehta 2010). Consequently, the energy cost pay-back time of these panels will be shortened to 0.5-2 years.

For the above mentioned reasons, a-Si/μc-Si solar panels are the mostly produced among all thin film technologies and will stay in large volume production the foreseeable future. This chapter introduces the fundamental thin film PV solar cell structure, the energy conversion physics, and state-of-the-art large scale solar panel manufacturing. Various methods of performance enhancement and cost reduction of large area thin film Si solar cells are focuses of this chapter.

This chapter is organized as follows. Section 1 briefly introduces the history and current production status of a-Si and μc-Si solar panels. Section 2 analyzes the cost structure of typical thin film solar panels and systems. The basic solar cell structures, including the PV active Si p-i-n junction layers and the front and back contact layers, are discussed in Section 3. Next, we describe in details the panel production process in Section 4 and 5. The front end of line (FEOL) processis first introduced, with discussions on CVD deposition of Si layers, physical vapor deposition (PVD) process of transparent conductive oxide (TCO) layers and back contacts, and laser scribing steps. The back end of line (BEOL) process is then described with the introduction of module fabrication, bus line wiring and panel encapsulation. Different process flow configurations are also compared in this part. We summary the chapter in Section 5.

2. Cost structure of PV system

To begin the discussion of the cost of solar panels, we split the cost of thin film PV system into four major parts:
1. Planning and financing: 15%
2. Inverter: 9-10%
3. Balance of system (BOS) and installation: 10-30%
4. Module: 40-66%

Sharing similar cost percentage of the first three parts with crystalline Si PV systems, the much lower module cost gives thin film PV system lower overall cost and a higher development potential. An increase or decrease of the efficiency of the module implies an increment or a reduction of the BOS and installation costs, respectively. Nevertheless, the financing and inverter cost remain always the same. Therefore, the use of lower efficiency thin film modules are financially more favorable in those cases in which the value of the installed area is not relevant. Thin film panels are thus more applicable to the PV electricity power plants built in remote areas like deserts. Large volume production and deployment is the key factor to fully demonstrate the financial benefit of thin film solar modules.

The cost of thin film modules, in turn is composed of five major components (Jäger-Waldau 2007):

1. Material cost (40%). The material consumption is determined by the film growth technology (e.g., PVD vs. CVD), and is also dominated by the module packaging and assembly technology. Special, TCO-coated glass substrates take a significant portion of the direct material cost (25-40%). Assuming similar technology used, the materials cost is inversely proportional to the production volume and panel efficiency.

2. Equipment related (capital) spending (20%). Initial investment on equipment on a-Si/μc-Si thin film panel manufactures is generally expensive. Upon fixed initial equipment investment, the annual depreciation of equipments is dominated by the deposition materials. The equipment depreciation rate is inversely proportional to the process throughput and module efficiency.

3. Labor cost (15-17%). The layered, monolithically integrated panel structure minimizes human operation, and the highly automated production methods used in the state-of-the art thin film PV panel manufactures reduce the labor cost. For a given total production volume, the labor cost is inversely proportional to the process throughput, extent of automation, and production module efficiency.

4. Energy consumption (15%). Modern PV manufactures use a significant amount of energy to run the factory, including machinery power consumption used for manipulating the substrate, controlling of substrate temperatures, RF power generators, film deposition system, vacuum system, exhaust handling, laser tool, lighting, air conditioning, etc. Once a factory is set up, a large amount of the overhead energy consumption is fixed, and the energy consumption per module is inversely proportional to the process throughput.

5. Freight (7-9%). The logistics of shipping and handling of the raw material as well as the assembled module panels take a larger portion of cost in thin film solar panels compared to their crystalline counter parts due to their greater size and weight. Unlike the other factors, freight cost is relatively constant for each panel.

As seen from the relationship of the thin film PV module cost structure summarized in Fig. 3, the process technology determines the direct material and energy consumption, equipment depreciation and ultimately the panel efficiency, which in turn affect the panel cost. In another word, more advanced module process technology leads to both higher panel efficiency and lower panel cost. Thus in this chapter, we put our focus on the process details of the manufacturing of modern, large-area a-Si/μc-Si solar panels.

3. Basic thin film Si solar cell structure

Typical a-Si single junction solar cells are composed of five principal layers: Si p-i-n diode sandwiched between two conductive layers. The front TCO forms the front contact, and the

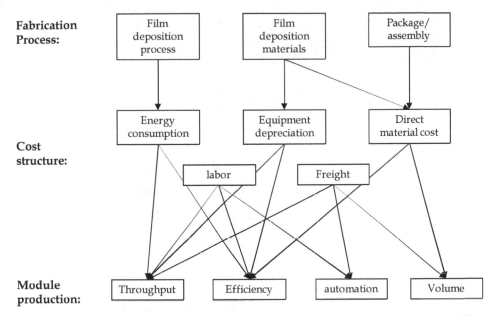

Fig. 3. Relationship between fabrication process, cost structure, and production of thin film PV modules.

back TCO and the reflector form the back contact (Fig. 4). The Si p-i-n junction absorbs sun light and generates photocarriers, which are collected by the conductive, front and back contacts. The substrate (e.g., glass) provides mechanical support for all the layers. Stacking two a-Si/μc-Si cells on top of each other forms the tandem junction structure, which is also sandwiched between the front and back TCOs.

Depends on the type of substrates on which the films are grown, there are basically two kinds of cell structures. 1) "Substrate" structure, where none-transparent substrates, i.e., metal foils, are used for growing the film stack. Sun light enters the cell from the top of the film stack by going through the top TCO. 2) "Superstrate" structure, where transparent substrates like glass or plastic films are used. Sun light enters the cell through the transparent glass/plastics and the TCO layer. The growth order of the Si p-i-n diodes are reversed in the two structures. The monolithically integrated superstrate type solar cells have superb encapsulation and compatibility with conventional electrical and safety regulations, thus holding a dominant market share.

3.1 PV active Si p-i-n layers

The Si p-i-n junction is where the sun light is absorbed and converted to charge carriers, i.e., electrons and holes. Differs from crystalline Si (c-Si), a-Si for PV and other applications (e.g., thin film transistor, TFT) are actually hydrogenated amorphous silicon alloy (a-Si:H, here noted as a-Si for simplicity), in which the H atoms passivate the otherwise high-density Si dangling bonds in pure amorphous Si film that introduce trap states and severely affect the film electrical properties. Normally the H content can be as high as a few percent. The a-Si completely loses the periodical atomic lattice structure; instead, the Si atoms randomly

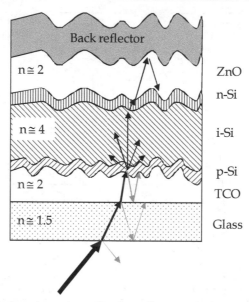

Fig. 4. Schematic single junction p-i-n a-Si solar cell. n stands for the index of refraction.

arrange in space. The lack of lattice structure makes a-Si a direct band gap semiconductor with a band gap of 1.8-1.9 eV at room temperature.

Hydrogenated microcrystalline Si (μc-Si:H, noted as μc-Si for simplicity) has a more complex, phase-mixed structure that consists of the crystalline phase made of silicon nanocrystallites and the amorphous Si matrix. The nanocrystallites grow into conglomerate clusters perpendicular to the film surface, whose diameters are typically between 10 and 50 nm. Embedded in amorphous silicon, the conglomerates are separated by a-Si, grain boundaries and micro-cracks. The band gap of μc-Si is 1.11 eV at room temperature, roughly the same as crystalline Si.

Photon absorption is proportional to the wavelength-dependent absorption coefficient, α, of the film. For typical a-Si and μc-Si, α is between 10^2 and 10^5 cm^{-1} in the visible range (Shah et al. 2004), which is 10-50 times larger than that of c-Si. Large α naturally allows for thinner absorber in solar cells. In the a-Si/μc-Si i-layer, an absorbed photon excites an electron from the valence band to the conduction band, creating a free electron and leaving a hole in the valence band. Due to the amorphous nature of a-Si and μc-Si films, the electrons and holes haves limited diffusion length and short life time. Electronic carrier transport properties are normally characterized by the mobility × lifetime product ($\mu\tau$-product), which is the physical characteristic of both carrier drift and diffusion processes. The measured products of the electron mobility and lifetime, $\mu^0\tau^0$, is 2×10^{-8} cm^2/Vs for a-Si and 1×10^{-7} cm^2/Vs for μc-Si, respectively, much lower than those measured in c-Si wafers (Beck et al. 1996; Droz et al. 2000). The low $\mu\tau$-product in a-Si or μc-Si makes the p-n diode configuration that is widely used in c-Si solar cells unsuitable with these materials, as the photocarrier collection in a p-n diode is diffusion limited. To avoid electron and hole recombination, p-i-n junction is used, where the built-in field drifts electrons towards the n-layer and holes towards the p-layer. The measured electron diffusion length is 2 μm in a-Si and 10 μm in μc-Si under the filed of

10^4 V/cm, comparable to or larger than the thickness of the solar cell film stack. As a result, the p-i-n type cells have efficient carrier collection efficiency.

Photons absorbed in the heavily doped n- and p- layers, however, don't contribute to the photocurrent as there is no net electric field in the doped layers. As a result, the n- and p-layers are usually less than 20 nm thick to limit photon absorption in these "window" layers. Further reduction of photon absorption in realized by increasing the band gaps of the n- and p-layers, e.g., doping the a-Si or μc-Si window layers with carbon so that they are transparent to the portion of the solar spectrum to be harvested in the i-layer. Total thickness of a typical single or tandem junction cell is less than 2 μm, which is only a few percent thick of a c-Si cell.

3.2 Front and back contacts

Though not PV active, the front and back contact layers play important roles on the cell performance. Optical wise, the transparent TCO layers scatter the incident sun light and enhance the optical absorption inside the i-layer. Electrical wise, since the lateral conductance of thin, doped p/n silicon layers is insufficient to prevent resistive losses, the TCO contact layers conduct the photocurrent in the lateral direction to the panel bus lines. The TCO layers used for thin film solar cells are doped wide band gap semiconducting oxides.

For efficient material usage and fast film deposition, the a-Si/μc-Si absorbers are so thin that the incoming light will not be completely absorbed during one single pass for normal incident rays. Hence, for all absorber materials, optical absorption inside the silicon layers has to be enhanced by increasing the optical absorption path. The difference of index of refraction between the TCO layers and the Si layers, plus the rough interface induce diffusive refraction of incoming light at oblique angles, thus increasing the optical path of solar radiation (Fig. 4). This is typically done by nano-texturing the front TCO electrode to a typical root-mean-square (rms) surface roughness of 40–150 nm and/or nano-textured back reflectors. In the ideal case, these rough layers can introduce nearly completely diffusive transmission or reflection of light (Müller et al. 2004).

When applied at the front contact, TCO has to possess a high transparency in the spectral region where the solar cell is operating (transmittance > 90% in 350 – 1000 nm), strong scattering of the incoming light, and a high electrical conductivity (sheet resistance < 20 Ω/sq.) (Fortunato et al. 2007). For the superstrate configuration where the Si layers are deposited onto a transparent substrate (e.g., glass) covered by TCO, it has to have at the same time favorable physicochemical properties for the growth of the silicon. For example, the TCO has to be inert to hydrogen-rich plasmas, and act as a good nucleation layer for the growth of the a-Si/μc-Si films. For all thin-film silicon solar cells, scattering at interfaces between neighboring layers with different refractive indices and subsequent trapping of the incident light within the silicon absorber layers is crucial to high efficiency.

TCO is also used between silicon and the metallic contact as a part of the back reflector to improve its optical properties and act as a dopant diffusion barrier. The back TCO layer also prevents reaction between the metal and the a-Si/μc-Si underlayers. Furthermore, applied in a-Si:H/μc-Si:H tandem solar cells, TCO can be used as an intermediate reflector between top and bottom cells to increase the current in the thin amorphous silicon top cell (Yamamoto et al. 2006). Finally, nano-rough TCO front contacts act as an efficient antireflection coating due to the refractive index grading at the TCO/Si interface.

The front and back TCO layers are at the same time electrodes that collect photogenerated carriers. As a semiconductor, the optical transparency and the electrical conductivity are closely related to the band gap structure of the TCO. The short-wavelength cutoff of the transmission spectrum corresponds to the oxide band gap, whereas the long-wavelength transmission edge corresponds to the free carrier plasma resonance frequency. On the other hand, electron conduction in TCO is achieved by degenerate doping that increases the free carrier density and moves the Fermi level into the conduction band. High carrier density and carrier mobility are thus required for TCO layers. There is, however, a tradeoff between high optical transmittance and low electrical resistance. Increasing electron carrier density decreases resistivity but also increases the plasma oscillation frequency of free carriers, thus shifting the IR absorption edge towards the visible. The transmission window is thus narrowed as a result of improved conductivity.

TCO type	ITO	ZnO:Al	SnO$_2$:F
Optical transmission (350-1000 nm)	95%	90-95%	90%
Resistivity ($\Omega \cdot$cm)	1-5×10^{-4}	3-8×10^{-4}	6-10×10^{-4}
Work function (eV)	4.7	4.5	4.8
Band gap (eV)	~3.7	~3.4	4.1-4.3
Deposition methods	RF sputtering	RF sputtering, LPCVD	APCVD, spray pyrolysis
Surface roughness	Flat	Excellent	Excellent
Plasma stability	Low	Excellent	Good
Relative cost	High	Middle	Low

Table 1. Different TCOs employed in Si thin film solar cells. RF, radio frequency. LPCVD, low-pressure chemical vapor deposition. APCVD, atmosphere-pressure CVD.

The most-widely used TCOs in Si thin film solar cells are doped SnO$_2$ (i.e., SnO$_2$:F) and ZnO (i.e., ZnO:Al) due to their temperature and chemical stabilities. Compared to the more conductive alternative indium-tin-oxide (ITO), they offer a much lower cost by avoiding the use of the costly In. At the same time, surface roughness induced by the crystalline texture of SnO$_2$ and ZnO is widely applied for increasing the optical absorption. These three typical TCOs are compared in Table 1.

The reflector layer on top of the back TCO can be Ag, Al, or white paint in a "superstrate" cell, and is the metal foil itself or another Ag/Al coating on the foil in a 'substrate' type cell. Ag is typically used in laboratory research work, while Al is more often used in mass production modules due to its lower cost and better properties in removing module shunts. Back contacts of Oerlikon's thin film panels, on the other hand, use proprietary white paint as the reflector (Meier et al. 2005). The white paint can be rolled on or screen printed directly onto the TCO. At the same time, it offers the following advantages (Berger et al. 2007): 1) high optical reflectance over a broad wavelength band, 2) optimal light scattering pattern which is generally beneficial for solar cells because this maximizes the fraction of photons that are trapped inside the solar cell due to total internal reflection at both cell surfaces, 3) pigmented materials have the potential of low cost. In certain instances, the white paint is a better surface reflector than Al, or TCO/Al reflector.

4. Factory panel production

As previously discussed (c.f., Section 2), manufacturing of the solar panels directly determines the cost of modules, which takes 40-66% of the overall PV system cost. This section focuses on the factory panel production, and addresses various methods of panel efficiency improvement and cost reduction.

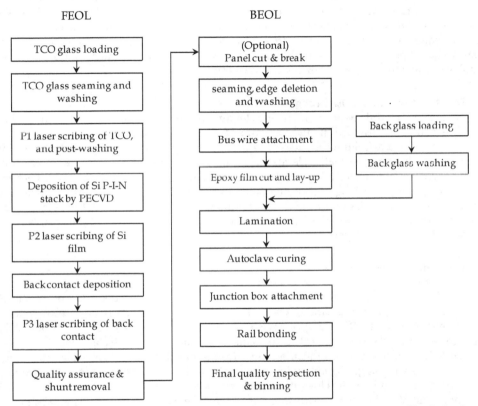

Fig. 5. Flow diagram of typical thin film solar panel production, comprising of both the front end of line (FEOL) and back end of line (BEOL) technologies and processes (Bhan et al. 2010).

Development in the a-Si and µc-Si thin film process technology combined with the booming PV market resulted in the fast expansion of a-Si/µc-Si based solar panel manufacturing after 2007. This industry largely benefits from the lab demonstration of thin film solar cells on small size substrates, as well as the large-area thin film deposition techniques developed for thin-film transistor liquid crystal display (TFT-LCD) industry. The growth of high-quality Si thin films for PV applications shares many of the skill sets required for growing Si TFT films, and using similar large-area thin film deposition chambers (Yang et al. 2007). In fact, both thin film solar "turnkey" equipment providers, Oerlikon and Applied Materials, have been manufacturing large-scale TFT-LCD deposition systems for years before becoming thin film solar equipment providers.

The general process flow of thin film panel fabrication is shown in the diagram of Fig. 5. This is a typical Applied Materials Sunfab process configuration (Bhan et al. 2010), though that of Oerlikon (Sun et al. 2009) and other 'superstrate' type panel makers are similar. The entire process is divided into the FEOL steps where the front glass substrate is deposited with active layers, and BEOL steps where the module is encapsulated. The FEOL mainly involves several film deposition and patterning steps, including the growth of Si p-i-n junctions by CVD and the growth of TCOs and metal layers by PVD. Laser scribing steps are used in between film depositions to form monolithically interconnected cells across the entire substrate. In the BEOL steps, the front glass with blanket film deposited during FEOL is cut, shaped and encapsulated to make solar panels. Optical and electrical inspections are taken upon finishing of FEOL for quality control purposes. Process of the 'substrate' type module shares many of the FEOL and BEOL steps, and the differences are discussed later in Section 4.3.

4.1 Front end of line (FEOL) process

The first step of module deposition involves loading of substrates in FEOL. Either float glass or TCO-coated glass is used, though an extra TCO growth step is required for the former case (Kroll et al. 2007; Sun et al. 2009). To prevent the glass from chipping and cracking during the following thermal cycle steps, as the film deposition steps require substrate temperatures ranging from 150 to 250 °C, the glass edge is seamed and reinforced. Then the glass goes through washing and drying steps to thoroughly remove debris, particles and organic contaminants. Laser-scribing step P1 follows to form isolated TCO contact strips. After a second washing step, the TCO glass is loaded into the film deposition chambers for the growth of PV active layers.

4.1.1 Growth of a-Si/μc-Si layers

The PV active Si p-i-n film stack is normally grown by CVD from gaseous precursors. Several CVD deposition techniques have been developed for the deposition of a-Si/μc-Si layers, including plasma-enhanced CVD (PECVD) (Schropp and Zeman 1998), remote plasma enhanced CVD (RPECVD) (Kessels et al. 2001), and hotwire CVD (Schroeder 2003). Though efficient lab-size solar cells are made with various CVD techniques, PECVD is prevailingly used for current industrial, high-throughput thin film PV module fabrication, as it possesses advantages like high deposition rate, in-situ chamber cleaning, good control over film quality, and requires lowest substrate temperature.

A typical PECVD chamber is structured like the schematics in Fig. 6. The substrate is supported by a susceptor, directly facing a gas diffuser. Process gases (SiH_4, H_2 and dopant gases) are fed into the chamber and dispersed by the diffuser. The diffuser and the susceptor are charged at opposite radio frequency (RF) voltages, thus exciting plasma within the chamber. Commonly used RF plasma is excited at 13.56 MHZ or 40 MHz, while higher RF frequencies are also used (Nishimiya et al. 2008). Higher RF frequencies are reported to deposit Si film faster due to higher plasmornic excitation energy. On the other hand, higher frequency means shorter RF wavelength, potentially forming standing wave inside the chamber that can cause non-uniform plasma distribution and reduce the a-Si/μc-Si film uniformity.

Using silane (SiH_4) as the precursor gas, the deposition of a-Si/μc-Si films can be described as a four-step process (Schropp and Zeman 1998):

1.   Primary gas phase SiH_4 decomposition. The plasma excites and decomposes the SiH_4 molecules into neutral radicals and molecules, positively and negatively charged ions, and electrons.

2. Secondary gas phase reaction. The reaction between molecules, ions and radicals (product of the previous step) generates reactive species and eventually large Si-H clusters, which are also called dust or power particles. The neutral species diffuse to the substrate, while the positive ions bombard the growing film and the negative ions are constrained within the plasma.
3. Film deposition. The radicals diffusing to the substrate interact with the substrate surface in various ways, like radical diffusion, chemical bonding, hydrogen sticking to the surface or desorption from the surface.
4. Formation of a-Si film. The actively growing film then releases hydrogen and relaxes into the Si network.

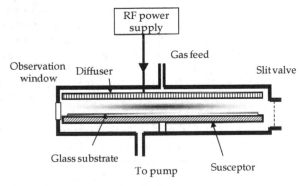

Fig. 6. Schematics of a PECVD process chamber.

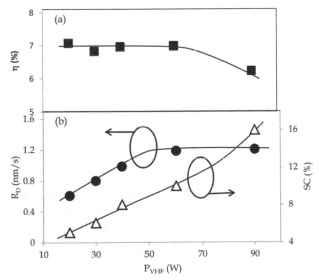

Fig. 7. Challenges met with high deposition rate (R_D) for PECVD grown μc-Si films. (a) Efficiencies of the optimal μc-Si single junction solar cell at different very-high frequency radio-frequency power (P_{VHF}). (b) Corresponding deposition rates RD and SiH$_4$ dilution concentration (SC) (Mai et al. 2005).

The overall film deposition is a complex process of gas and surface reactions, and is controlled by many deposition parameters, including gas composition, flow rate, chamber pressure, RF power density, RF frequency, substrate temperature, and the chamber size and geometry. Extensive studies are carried out to study the influence of those controlling parameters on the a-Si/μc-Si film properties, and are summarized in review papers (Luft and Tsuo 1993; Bruno et al. 1995). Perusing high solar energy conversion efficiency requires high quality, PECVD grown a-Si/μc-Si films, like high optical absorption, low dangling bond density, wide band gap for optical transmission of the p- and n- window layers (Schropp 2006).

Transferring the lab-developed process to solar panel production, however, has its own challenges. The PECVD steps take a significant portion of the cost in energy consumption and equipment depreciation (c.f., Fig. 3), thus substantial changes must be made to the lab-developed PECVD processes in panel manufacturing to fit the goal of lowering the panel cost. While maintaining the optimal panel performance, the widely adapted strategies are high rate of deposition (R_D) and large substrate size for high-throughput panel growth.

As previously discussed in Section 2, fast film deposition can lower the energy consumption, facilitating the throughput and process efficiency, thus effectively lower the cost of solar modules. The major obstacle to throughput increase is the μc-Si deposition, which has to be thick (> 1.5 μm) as limited by the finite absorption coefficient, thus requiring appreciably long deposition time. In fact, deposition of the μc-Si bottom cell in a-Si/μc-Si tandem junction solar panels takes the longest process time in the Oerlikon and Applied Materials process lines. To shorten the μc-Si deposition time and improve the overall throughput, research has been focused on increasing the deposition rate of μc-Si films.

Generally, increasing the density of SiH radicals promotes the growth of a-Si/μc-Si, thus increasing SiH_4 flow rate, applying higher RF power density, or using a higher RF excitation frequency all lead to higher R_D. For the simpler case of a-Si, higher RF power and higher gas flow rate result in faster film growth. As for the case of μc-Si, changing these deposition parameters at the same time affects the film crystallization in addition to increasing R_D. Since a good performed μc-Si solar cell needs to keep at the transition from microcrystalline to amorphous growth, increasing the film deposition rate should not be compromised by the film crystallinity. It is observed that increase of RF power requires higher SiH_4 concentration to keep the same crystallinity, which at the same time leads to higher deposition rate. In one example, Fig. 7 compared the high R_D achieved with increasing very high frequency (VHF) 94.7 MHz RF power (Mai et al. 2005). To maintain the maximum PV efficiency (Fig. 7a) at different VHF power (P_{VHF}), the silence concentration (SC) in the mixed gas has to increase with P_{VHF} (Fig. 7b). At small P_{VHF}, R_D increases linearly with P_{VHF}, and saturates around 1.2 nm/s when P_{VHF} is > 60 W. The solar cell efficiency, η stays constant for the same region since the μc-Si crystallinity remains unchanged. Further increase of P_{VHF} doesn't lead to higher RD, and the optimal cell efficiency drops at this region. It is important to note that higher deposition rate greatly improves the throughput of the panel manufacturing, and lowers the panel cost.

Since the merit of solar cell, \$/Wp, is inversely proportional to the total solar module production (Hegedus and Luque 2003), growing solar cells over large-area substrates is one of the most efficient ways of lowering solar cell cost. For constant direct materials and labor cost, growing films over larger area substrates effectively lowers the module cost per unit area. The major challenge when scaling up the substrate is to maintain uniform film growth over large area. Since various PECVD parameters directly affect the growth rate of a-Si / µc-Si film, the non-uniform distribution of these growth parameters induces local film thickness variation. For typical p-i-n type a-Si or µc-Si cells, the open circuit voltage (V_{OC}) and fill factor (FF) decrease upon increasing i-layer thickness, while J_{SC} increases as a result of the higher absorption in the thicker cells (Klein et al. 2002). Other than thickness, the RF power distribution affects the crystallinity of the as-grown µc-Si, which in turn changes the solar cell performance. For example, slight deviation of RF intensity resulted in unbalanced µc-Si crystallization in a Gen 8.5 PECVD chamber, as shown by the smaller fraction of crystallinity (FC) on the left side of chamber before adjusting the RF power supply feed (Fig. 8a), though such deviation could be too small to affect the a-Si and µc-Si film thicknesses (Yang et al. 2009). Small size a-Si/µc-Si tandem junction solar cells cut from the solar panels at corresponding locations, had non-uniform performance distribution. Sample cells from the low RF intensity side (left) had smaller short circuit current density (J_{SC}) (Fig. 8b) and higher V_{OC} (Fig. 8c) than those from the other side, while FF had a uniform distribution despite the RF influence (Fig. 8d). After modifying the RF feed of the PECVD chamber, balanced FC distribution was obtained, and the sample cells showed uniform distribution of J_{SC}, V_{OC} and FF. It is thus important to keep the uniform distribution of all process parameters in large scale process, and special attention must be paid to the RF power distribution across the whole chamber.

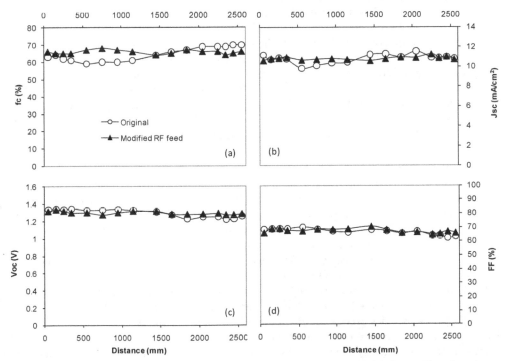

Fig. 8. µc-Si film uniformity and its impact to solar cell performance (a) Crystalline fraction (fc) change along substrate diagonal with original and modified RF feed. (b) J_{SC}, (c) V_{OC}, and (d) FF profiles along substrate diagonals. The two substrates were grown before and after modifying the RF feed location.

4.1.2 Growth of TCO layers

The TCO layers are typically grown by either CVD or PVD on industrial size substrates. Used as the front TCO layer, SnO_2:F layer are normally grown by CVD using $SnCl_4$ precursor gas on glass substrates (Rath et al. 2010). The finished SnO_2:F-coated glass is provided as substrates for later PECVD deposition of a-Si/μc-Si layers. Another front TCO candidate is AZO, which is usually grown by CVD processes using precursors like $Zn(C_2H_5)_2$ or sputter deposition (Agashe et al. 2004). Substrate temperatures are near 150 °C for the CVD deposition, and can go as high as 300 °C for sputtering.

For many solar panel manufactures, control of the front TCO properties, especially the surface roughness, can be financially unfeasible, thus purchasing glass substrate pre-coated with TCO becomes a good choice. For a-Si absorber layers a high transparency for visible light (wavelength λ = 400–750 nm) is sufficient, while for solar cells incorporating μc-Si the TCO has to be highly transparent up to the near infrared (NIR) region (400–1100 nm) to accommodate for the wider absorption spectrum of μc-Si. This imposes certain restrictions on the carrier density, n, of the TCO material, as increased free carrier absorption leads to a reduction of IR transmission (Agashe et al. 2004). SnO_2:F films fulfilling these requirements to a large extent, have been developed by Asahi Glass (Asahi Type U) (Sato et al. 1992). The ZnO crystallite facets imposes diffusive light scattering to the incident sun light, thus enhances the optical absorption with minimal absorbing layer. Further light trapping for long-wavelength light is also achieved in the Asahi high haze, HU-TCO glass, where the TCO surface has two types of textures of different characteristic length thus scattering different portions of light (Kambe et al. 2009).

The back TCO is typically AZO (c.f., Table 1), which is grown in-line with other in-house process steps. In addition, AZO provides excellent long-term stability as the back contact material. Since the back contact is grown on top of finished Si p-i-n stack, the substrate temperature should be kept low (<300 °C) to prevent the dopant diffusion from the a-Si/μc-Si n layers. The back AZO contact is typically grown by RF sputtering (Beyer et al. 2007) or low pressure (LP)-CVD (Meier et al. 2010) at temperatures < 150 °C.

4.1.3 Laser scribing

A big advantage of the superstrate type a-Si/μc-Si thin film solar panels lies in the monolithically integrated structure, which greatly reduces operational cost and increases production throughput and panel yield by eliminating connection of wafers in the fabrication of crystalline Si PV panels. The thin film panel is scribed into numerous small cells, which are interconnected in series for a high voltage output, which at the same time improves panel yield and lowers the resistive energy loss. The monolithically integrated series connection is realized by a three step patterning process that selectively removes the individual layers, i.e., TCO front contact, thin-film silicon layer stack, and back contact of the solar cell. Highly automated laser scribing patterning is widely used for all patterning steps (c.f., Fig. 5) as it provides precise positioning, high throughput and minimum area losses.

Scribing thin film layers into sections with laser has developed a mature technology applied to large-area superstrate-type module fabrication, where laser beam incidents through the glass substrate. As depicted in Fig. 9, laser scribing mechanism can be described as a four-step process (Shinohara et al. 2006).

1.   Absorption of laser beam. By choosing appropriate laser energy/wavelength, the layer to be scribed absorbs the laser energy following Lambert's law, with more heat created

at the glass-side of the film (Fig. 9a). Typical laser energy is $>1\times10^6$ W/cm^2, and calculation shows more than 80% of that energy is absorbed and converted to heat building up in the film.

2. Decomposition of H from a-Si:H. The absorbed heat induces the decomposition of a-Si:H, and releases hydrogen at a temperature of > 600 °C (Fig. 9b). In fact, the local temperature in the film can be heated up to 700 °C by the laser.

3. Destruction of the PV layers and back contact. The gaseous H$_2$ quickly expends its volume and pressure under the high temperature. The pressure of the H$_2$ gas can amount to $>1\times10^7$ Pa, inducing enormous shear stress on the layers above the heated zone. In one estimation, applying a 532 nm, 12 kHz and 9.5×10^6 W/cm^2 laser beam on a-Si single junction module created shear stress of 3.9×10^8 Pa, enough to break the layers on top of the heating zone, among which the most ductile Ag layer has a shear strength of $10^7 - 10^8$ Pa (Fig. 9c).

4. Formation of heat affected zone (HAZ). Along with the H$_2$ volume expansion, the film cracks quickly followed by blasting off, effectively removing the a-Si/μc-Si layers and the back contact layers above the local, heated zone. The laser heating also damages the film around the removed region, creating a HAZ with high density of defects and poor electrical properties (Fig. 9d). By using high-frequency pulsed laser, the HAZ is limited to less than a few tens of nm wide.

It is important to note that the laser scribing removal is not a true thermal process but the mechanical blasting off of the film. By applying different wavelengths of lasers, the laser energy is absorbed by different layers, thus selectively removes those layers without affecting other, underlying layers.

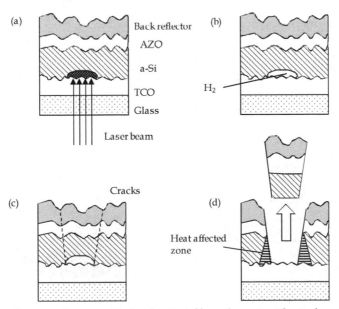

Fig. 9. Laser-scribing mechanism. (a) absorption of laser beam incidents through the glass, (b) decomposition of H from a-Si:H, (c) destruction of the photovoltaic layer and back electrode, (d) film blasted off and formation of heat affected zone (Shinohara et al. 2006).

Combination of several laser-scribed layers is used to create interconnection in Si thin-film modules (Fig. 10a). The cell strips are defined by selective ablation of individual layer stacks, and the interconnection between neighboring strip cells are provided by the overlap of conductive layers. In the microscope view of a typical interconnection area (Fig. 10b), P1 is the first laser scribing step that cuts through the front TCO layer, P2 is the second scribing step that cuts through the p-i-n junction layers, P3 is the last step that cuts through the junction layers and the back reflector. The dead-area, i.e., the narrow area between P1 and P3 lines including the HAZ, makes up the interconnection junction but doesn't contribute to photocurrent generation. State-of-the-art laser process can limit the interconnection width to < 350 μm to minimize the dead-area. For a-Si/μc-Si module production, the scribing laser is typically powerful Nd:YVO$_4$ solid-state laser with primary emission at 1064 nm and second harmonic generation at 532 nm. P1 is scribed by the 1064 nm irradiation, in which the strong absorption in TCO results in intensive local heating and explosive TCO evaporation (ablation); the glass that doesn't absorb in this wavelength keeps cool and is free from damage. P2 and P3 are similarly scribed by the 532 nm irradiation. As shown in Fig. 10c, the P3 laser cuts abrupt edges on the a-Si film without leaving any observable damage to the underlying TCO layer. The three laser scribing steps combining the subsequent film deposition steps form differences in the depths of different layers and conductive channels, forming the interconnection region of the cell strips' series connection. Power optimized, high-speed laser scribing technique is already applied in making 5.7 m^2 solar panels with exceptional performance (Borrajo et al. 2009).

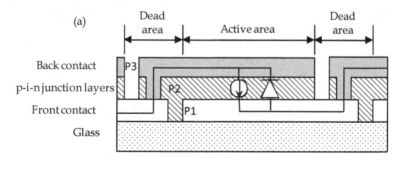

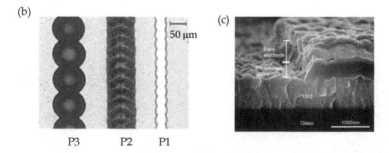

Fig. 10. (a) Schematic cross-sectional view of Si thin film solar panel showing the sectioned film and laser scribing lines (P1, P2 and P3). (b) Optical microscope image of the laser scribed lines. (c) Scanning electron microscope image of P3 scribed PV layer (Shinohara et al. 2006).

Making up the interconnection of cell strips, the laser scribing pattern is decisive to the assembled panel performance. Since the total panel area is fixed, the width of cell strips determined by the laser scribing pattern is inversely proportional to the number of cell strips. Laser scribing pattern also affects the number of junctions and total dead area, which both contribute to losses in panel power output. Thus design of the laser scribing pattern is optimized with the width of strip cells and the sheet resistance of the front and back contacts.

Precisely scribing fine lines that defines the monolithically integrated thin film solar module, laser scribing technology greatly enhanced the overall panel performance and improves the automation of the process flow. It is an important step in improving the module efficiency and driving down the module cost independent of the film deposition processes.

4.1.4 Rest of FEOL steps

After all film deposition and laser scribing steps conclude, the central PV active region is isolated from the panel edge to avoid electrical shock. In one way, the outside edge of the entire film stack is removed by 10-20 mm of width, called edge deletion. This is typically done by mechanical grinding or laser scribing (same as P2 laser).

To burn out the defects and improve panel yield, the final FEOL step involves removing cell shunts by reverse biasing the cells, or shunt busting. Shunting in Si thin film solar cells refers to high leakage current in reverse bias, which leads to a loss of power and efficiency. In large scale deposition, pinholes or locally thinner Si layer could form, which allow a connection between the top and bottom contacts, forming partially shorted PV diodes. When applying a reverse bias, larger current is focused at these shunt regions, resulting in local heat generation and consequent burning out of the low resistance pathway. Microscopic observation confirms the change of film morphology and its connection to the curing of the solar cells (Johnson et al. 2003).

As all cells are readily formed at this stage, electrical and optical inspection of individual cell strips are taken after the shunt busting for quality assurance purposes. This completes the FEOL processing of the solar panel.

4.2 Back end of line (BEOL) process

Panels fabricated at FEOL have to be further shaped and encapsulated to complete the solar panel module at the BEOL steps. Though no more film is deposited in the BEOL steps, these are important processes to ensure high quality solar panel production.

4.2.1 Module fabrication and bus line wiring

If the module size is smaller than the substrate, glass with deposited film is first scored and broken into the final panel size, and goes through edge deletion.   Then the panel is thoroughly washed for another time and ready for final bus line soldering.

According to the laser scribing layout, the two terminal segments of the series connected cell strips are each soldered to a bus line. These two terminal segments serve as the beginning and ending of the series connection of all cell strips on the panel. The cross bus bars are then attached to the terminal bus line and leaves out the final electrical connection to the external circuit.

4.2.2 Module encapsulation

To stand for extreme weather conditions in field usage, the functional films, i.e., TCO layers, a-Si/µc-Si films, metal coatings, and bus lines need good encapsulation to achieve for long panel lifetime. The most common encapsulation method for panels with the glass substrate is to use another piece of glass to cover the functional films. The gap between the two glass plates is filled with an epoxy (ethylene vinyl acetate, EVA, or polyvinyl butyral, PVB) film, which not only insulates the functional films against reactants like oxygen and moisture, but also mechanically strengthens the rigidness of the finished panel. Quality of the module encapsulation is directly associated with the failures of panels in the field. Judgment of the encapsulation properties includes low-interface conductivity, adequate adhesion of encapsulants to glass as a function of in-service exposure conditions, and low moisture permeation at all operation temperatures (Jorgensen et al. 2006).

The panel then passes through a laminator where a combination of heated nip rollers removes the air and seals the edges. The lamination film at the same time provides electrical insulation against any electric shock hazard. At the exit of the laminator conveyer, the modules are collected and stacked together on a rack for batch processing through the autoclave where they are subjected to an anneal/pressure cycle to remove the residual air and completely cure the epoxy. Finally, a junction box is attached to the cross bus wire and sealed on top of the hole of the back glass and is filled with the pottant to achieve a complete module integrity.

The fully processed module is then tested for output power, I_{SC}, V_{OC}, and other characteristics under a solar simulator. Then it is labeled, glued to the supporting bars, and packaged. At this point, the full panel assembly is finished.

4.3 Production process flow

Multiple chambers are used for deposition of different functional layers in the module production process. Optimizing the arrangement of chambers and controlling of the process flow are crucial to the production throughput and directly affect the panel production cost. There are mainly three types of process flows: batch process, continuous process and hybrid process. Characteristics of the three processes are compared in Table 2.

Batch process of film deposition is the most intuitive way of arranging deposition chambers. In this configuration, functional layers are deposited consequently onto batches of substrates. The typical batch processes are seen in Oerlikon's thin film production lines. An example is the Oerlikon KAI-20 1200 production system (Fig. 11a), which consists of two PECVD process towers, two load-locks, one transfer chamber and an external robot for glass loading from cassettes (Kroll et al. 2007). Each process tower is equipped with a stack of ten plasma-box-reactors where ten substrates are deposited simultaneously. The layers are processed in parallel at the same time in both stacks (2×10 reactors). The whole KAI-20 1200 PECVD production system shares one common gas delivery system including the mass flow controllers and one common process pump system. Engineering work has been put to ensure small box-to-box variations of deposition rates, layer thickness uniformities. The batch process normally requires small footprint, and is suitable for slow deposition that requires long process time (e.g., the absorbing i-layers). In fact the PECVD deposition of different p-, i-, and n- layers can be combined within the same chamber as long as dopant diffusion from the process chamber can be minimized. In most cases more than one chambers are used for the entire film stack, thus when they are moved between separate chambers the substrate manipulation and heating / cooling time has to be minimized to increase the process throughput.

Process flow types	Batch Process	Continuous Process	Hybrid Process
Schematics	Fig. 12a	Fig. 12b	Fig. 12c
Examples	Oerlikon Solar customers	United Solar, ECD, Xunlight	Applied Materials customers
Production volume	20 MW/yr	30 MW/yr	40-55 MW/yr
System footprint	6 m × 8.6 m (KAI 1200)	6 m × 90 m	Variable sizes
Substrate	Glass, 1.1 m × 1.25 m	Stainless steel roll, 36 cm x 2.6 km	Glass, 2.2 m × 2.6 m
Operational flexibility	Same equipment can be used for multiple depositions	Moderate operational flexibility but often leads to inefficient capital use.	Same equipment can be used for multiple depositions
Standardized equipment	Easily modified to produce different solar cell structures	Recipe of the entire line is fixed. Equipments are optimized for minimal operating conditions	Easily modified to produce different solar cell structures.
Rate of deposition affects throughput	Favors slow depositions that require long residence time.	Slow depositions require large equipments and slow process flow.	Slow process is shared by parallel chambers for high throughput.
Processing efficiency	Requires strict scheduling and control. Minimal energy integration.	Reduces fugitive energy losses by avoiding multiple heating and cooling cycles	Scheduling and synchronization of chambers are optimized by artificial intelligence.
Product demand	Changing demand for products can be easily accommodated. Possible of making multiple solar panels with different structures.	Difficult to make changes as the process recipes are fixed for the entire line.	Changing demand for products can be easily accommodated. Possible of making multiple solar panels with different structures at the same time.
Equipment fouling	Tolerable to significant equipment fouling because cleaning / fixing of equipment is a standard operating procedure. Throughput can be affected when individual plasma-box fails in the process tower.	Significant fouling in continuous operations is a serious problem and difficult to handle. Sometimes significant fouling requires shutting down of the entire production line.	Fouling chamber can be by-passed or replaced with similar chambers, thus minimizing the adverse effect to the throughput.

Table 2. Comparison of three thin film solar module process flow types

Continuous deposition of the multilayer structure is realized in a roll-to-roll manner, which ensures stable chamber conditions for consistent film growth for large volume production. United Solar, Energy Conversion Devices (ECD), and Xunlight took this type of growth configuration. For example, the ECD 30 MW a-Si process line consists of nine series-connected chambers with gas gates that isolate dopant gases between chambers (Fig. 11b)

(Izu and Ellison 2003). The film deposition substrates are 2.6 km long, 36 cm wide, 127 μm thick stainless-steel rolls fed into the deposition system at constant speed. For quality assurance, online diagnostic systems are installed allowing for continuous monitoring of the layer thickness and characterization of the PV properties of the manufactured solar cells. A big advantage of the continuous process is that the substrate does not see the atmosphere during the process, and needs to be heated and cooled only at the beginning and last chamber, thus greatly saving the pumping time and energy cost. At the same time, all chambers continuously run at the optimized, stable states, thus depositing films with uniform and consistent properties. On the other hand, Since the deposition rate and thickness of each layer varies a lot (e.g., typical p-layers are < 20 nm while the μc-i layer is normally 1-2 μm), the deposition time in each chamber are very different. Limited by a constant substrate roll feeding speed, the chamber for growing i-layers are much longer than the doped layer chamber. In fact, this 30 MW system is 90 m long.

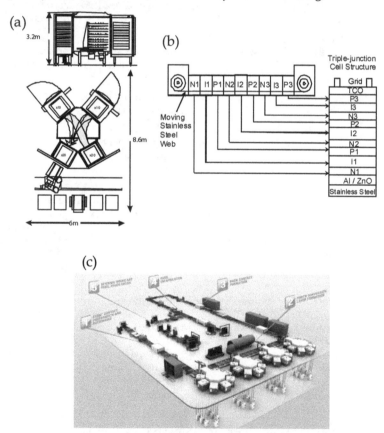

Fig. 11. Typical process systems used for Si thin film solar cell manufacturing. a) Batch process. Schematic side and top view of an Oerlikon KAI-20 1200 PECVD process system for a-Si deposition (Kroll et al. 2007). b) Linear Process. Schematic diagram of a United Solar Ovonic Corporation roll-to-roll a-Si:H alloy triple-junction solar cell processor (Yang et al. 2005). c) Hybrid (batch plus linear) process. Schematics of a Applied Materials SunFab thin film production line (Applied Materials 2010).

The hybrid-process system is designed to combine the advantages of batch and continuous processes. In this configuration, separated chambers are used like those in batch process, but individual substrates are fed into different chambers for optimal chamber utilization. Each substrate sees a queue of different process chambers like that in continuous process. Applied Materials configured its SunFab in the hybrid mode, where a group of several process chambers construct a functional cluster unit sharing a heating chamber and a center transfer robot (Fig. 11c). Each cluster is focused on a group of related functional layers (e.g., layers comprising a subcell in a multi-junction structure), and deposition of the multi-junction stack is realized by going through clusters. In this configuration, each chamber can have flexible deposition time, and the flow of substrates and synchronization of chambers are controlled by artificial intelligence algorithm for optimal system throughput (Applied Materials 2010; Bourzac 2010). This process flow combines the advantages of small footprint, easy maintenance and high production throughput, and provides flexible system configuration for versatile panel fabrication.

There are a number of considerations to weigh when deciding among batch, continuous or hybrid processes, and some of the major reasons are listed in Table 2. Generally, small production volumes favor the batch process type while continuous process is more suitable for high volume production. Capital investment cost of a batch or hybrid process system is also usually lower than the continuous process because the same equipment can be used for multiple unit operations and can be reconfigured easily for a wide variety of panel structures, though the operating labor costs and utility costs tend to be high for the former two systems (Turton et al. 2008). The continuous configuration is also more favored for 'substrate' type solar cells on metal foil substrates in a roll-to-roll deposition (Izu and Ellison 2003). Though the comparisons in Table 2 generally holds true, it is also possible that the configuration works for one solar plant may not be the best choice of another, as each plant differs at production scale, materials supply, geological confinement and many other practical characters.

5. Conclusion

In this chapter, the cost structures of a-Si/µc-Si solar modules has been described with analysis of the multilayer cell structure and module production. The monolithically integrated structure is described with explanations of layer functions. The industrial fabrication of large-area modules are introduced, including FEOL and BEOL process steps.

Module costs around half of the total thin film PV system. We analyzed the factors affecting the module efficiency and cost in terms of energy consumption, equipment investment, spending on direct material, labor and freight cost. To probe strategies of efficiency improvement, we started from the introduction of the Si p-i-n junction structure and the front/back contacts, and discussed the light absorption and its enhancement with light trapping. The photocurrent generation is achieved by effective capture of the incident solar photons, and conversion into free electrons and holes by the build-in field of the p-i-n junction. Resistance loss during photocurrent collections is minimized by the conductive front and back contact layers. At the meantime, enhancing the light absorption within thin layers is achieved using band gap engineering of the absorbing layer and optical trapping of the front/back contact layers.

Fabrication of large-area tin film solar panels are the key to increasing the production volume and reducing the $/Wp of modules. State-of-the-art fabrication includes FEOL and

BEOL process steps. In the FEOL processes, glass substrates are subsequently coated with functional layers, i.e., the a-Si/μc-Si layers by PECVD, TCO and reflector layers are grown by PVD or CVD. The monolithically integrated module structure is achieved by laser scribing of individual layers. In the BEOL processes, the panels are cut and encapsulated. Electrical wiring are also finished in the BEOL steps. The batch, linear, and hybrid process flow schemes are compared with actual factory examples.

Thin film a-Si/μc-Si solar panels have been holding the largest market share among all produced thin film panels. The power conversion efficiency of these panels is likely to increase to above 12% in the near future, but not exceed that achieved in crystalline cells. Advantages such as large-area, low-cost fabrication, and demonstrated field performance, nevertheless, render a-Si/μc-Si thin film technology attractive for large-area deployment like in solar power plants. In particular with the uncertain elemental supply becomes an issue for CdTe and CIS cells that might impair the sustainability of those PV products (Fthenakis 2009), thin film a-Si/μc-Si is likely to have long-term potential for providing energy supply in an even larger scale. Improvements on efficiency and stability would continue to drive the research in this area, while panel manufacturing will continue to be optimized for achieving lower production cost and optimal $/Wp.

6. References

Agashe, C., et al. (2004). Efforts to improve carrier mobility in radio frequency sputtered aluminum doped zinc oxide films. *Journal of Applied Physics* Vol. 95, No. 4: pp. 1911-1917, ISSN 2158-3226

Applied Materials, Inc. (2010). Applied Materials Solar, Last accessed 2011, Available from http://www.appliedmaterials.com/technologies/solar

Beck, N., et al. (1996). Mobility lifetime product---A tool for correlating a-Si:H film properties and solar cell performances. *Journal of Applied Physics* Vol. 79, No. 12: pp. 9361-9368, ISSN 2158-3226

Berger, O., et al. (2007). Commercial white paint as back surface reflector for thin-film solar cells. *Solar Energy Materials and Solar Cells* Vol. 91, No. 13: pp. 1215-1221, ISSN 0927-0248

Beyer, W., et al. (2007). Transparent conducting oxide films for thin film silicon photovoltaics. *Thin Solid Films* Vol. 516, No. 2-4: pp. 147-154, ISSN 0040-6090

Bhan, M. K., et al. (2010). Scaling single-junction a-Si thin-film PV technology to the next level. *Photovoltaics International* Vol. 7: pp. 101-106, ISSN 1757-1197

Borrajo, J. P., et al. (2009). Laser scribing of very large 2,6m x 2,2m a-Si: H thin film photovoltaic modules. *Processings of Spanish Conference on Electron Devices, 2009*, pp. 402-405, 11-13 Feb. 2009

Bourzac, K. (2010). Scaling Up Solar Power. *Technology Review* Vol. 113, No. 2: pp. 84-86, ISSN 1099274X

Bruno, G., et al. (1995). *Plasma Deposition of Amorphous Silicon-Based Materials (Plasma-Materials Interactions)*. Academic Press ISBN 978-0121379407

Droz, C., et al. (2000). Electronic transport in hydrogenated microcrystalline silicon: similarities with amorphous silicon. *Journal of Non-Crystalline Solids* Vol. 266-269, No. Part 1: pp. 319-324, ISSN 0022-3093

Fortunato, E., et al. (2007). Transparent Conducting Oxides for Photovoltaics. *MRS BULLETIN* Vol. 32, No. 3: pp. 242-247, ISSN 0883-7694

Fthenakis, V. (2009). Sustainability of photovoltaics: The case for thin-film solar cells. *Renewable and Sustainable Energy Reviews* Vol. 13, No. 9: pp. 2746-2750, ISSN 1364-0321

Green, M. A. (2007). *Third Generation Photovoltaics: Advanced Solar Energy Conversion* Springer, ISBN 978-3540265627, New York

Hegedus, S. S. & Luque, A., Eds. (2003). *Status, Trends, Challenges and the Bright Future of Solar Electricity from Photovoltaics.* John Wiley & Sons Inc., ISBN 0-471-49196-9, Chippenham, Great Britain

Izu, M. & Ellison, T. (2003). Roll-to-roll manufacturing of amorphous silicon alloy solar cells with in situ cell performance diagnostics. *Solar Energy Materials and Solar Cells* Vol. 78, No. 1-4: pp. 613-626, ISSN 0927-0248

Jäger-Waldau, A. (2007). Status and Perspectives of Thin Film Solar Cell Production. *Processings of 3rd International Photovoltaic Industry Workshop on Thin Films,* EC JRC Ispra, November 22-23, 2007

Johnson, T. R., et al. (2003). Investigation of the Causes and Variation of Leakage Currents in Amorphous Silicon P-I-N Diodes. *Materials Research Society Symposium Proceedings* Vol. 762: pp. A7.7.1-A7.7.6, ISSN 0272-9172

Jorgensen, G. J., et al. (2006). Moisture transport, adhesion, and corrosion protection of PV module packaging materials. *Solar Energy Materials and Solar Cells* Vol. 90, No. 16: pp. 2739-2775, ISSN 0927-0248

Kambe, M., et al. (2009). Improved light-trapping effect in a-Si:H / µc-Si:H tandem solar cells by using high haze SnO_2:F thin films. *Processings of Photovoltaic Specialists Conference (PVSC), 2009 34th IEEE,* pp. 001663-001666, Philadelphia, USA, June 2009

Kessels, W. M. M., et al. (2001). Hydrogenated amorphous silicon deposited at very high growth rates by an expanding Ar-H2-SiH4 plasma. *Journal of Applied Physics* Vol. 89, No. 4: pp. 2404-2413, ISSN 2158-3226

Klein, S., et al. (2002). High Efficiency Thin Film Solar Cells with Intrinsic Microcrystalline Silicon Prepared by Hot Wire CVD. *Materials Research Society Symposia Proceedings* Vol. 715: pp. A26.22, ISSN 0272-9172

Kolodziej, A. (2004). Staebler-Wronski effect in amorphous silicon and its alloys. *Opto-Electronics Review* Vol. 12, No. 1: pp. 21-32, ISSN 1230-3402

Kroll, U., et al. (2007). Status of thin film silicon PV developments at Oerlikon solar. *Processings of 22nd European Photovoltaic Solar Energy Conference,* pp. 1795-1800, Milan, Italy

Luft, W. & Tsuo, Y. S. (1993). *Hydrogenated amorphous silicon alloy deposition processes.* CRC Press, ISBN 978-0824791469, New York

Mai, Y., et al. (2005). Microcrystalline silicon solar cells deposited at high rates. *Journal of Applied Physics* Vol. 97, No. 11: pp. 114913-114912, ISSN 2158-3226

Mehta, S. (2010). Thin film 2010: market outlook to 2015

Meier, J., et al. (2010). From R&D to Large-Area Modules at Oerlikon Solar. *Materials Research Society Symposia Proceedings* Vol. 1245: pp. 1245-A01-02, ISSN 0272-9172

Meier, J., et al. (2007). UP-scaling process of thin film silicon solar cells and modules in industrial pecvd kai systems. *Conference Record of the 2006 IEEE 4th World Conference on Photovoltaic Energy Conversion, WCPEC-4,* pp. 1720-1723

Meier, J., et al. (2005). Progress in up-scaling of thin film silicon solar cells by large-area PECVD KAI systems. *Conference Record of the Thirty-first IEEE Photovoltaic Specialists Conference, 2005*, pp. 1464-1467

Müller, J., et al. (2004). TCO and light trapping in silicon thin film solar cells. *Solar Energy* Vol. 77, No. 6: pp. 917-930, ISSN 0038-092X

Nishimiya, T., et al. (2008). Large area VHF plasma production by a balanced power feeding method. *Thin Solid Films* Vol. 516, No. 13: pp. 4430-4434, ISSN 0040-6090

Rath, J. K., et al. (2010). Transparent conducting oxide layers for thin film silicon solar cells. *Thin Solid Films* Vol. 518, No. 24 SUPPL.: pp. e129-e135, ISSN 0040-6090

Repmann, T., et al. (2007). Production equipment for large area deposition of amorphous and microcrystalline silicon thin-film solar cells. *Conference Record of the 2006 IEEE 4th World Conference on Photovoltaic Energy Conversion, WCPEC-4*, pp. 1724-1727

Sato, K., et al. (1992). Highly textured SnO_2:F TCO films for a-Si solar cells. *Reports of the Research Laboratory, Asahi Glass Co., Ltd.* Vol. 42, No.: pp. 129-137, ISSN 0004-4210

Schroeder, B. (2003). Status report: Solar cell related research and development using amorphous and microcrystalline silicon deposited by HW(Cat)CVD. *Thin Solid Films* Vol. 430, No. 1-2: pp. 1-6, ISSN 0040-6090

Schropp, R. E. I. (2006). *Amorphous (Protocrystalline) and Microcrystalline Thin Film Silicon Solar Cells*. Elsevier B. V., ISBN 9780444528445

Schropp, R. E. I. & Zeman, M. (1998). *Amorphous and Microcrystalline Silicon Solar Cells*. Kluwer Academic Publishers, ISBN 978-0792383178 Boston

Shah, A. V., et al. (2004). Thin-film silicon solar cell technology. *Progress in Photovoltaics: Research and Applications* Vol. 12, No. 2-3: pp. 113-142, ISSN 1099-159X

Shinohara, W., et al. (2006). Applications of laser patterning to fabricate innovative thin-film silicon solar cells. *Processings of SPIE*, Vol. 6107, pp. 61070J-1-18

Sun, H., et al. (2009). End-To-End Turn-Key Large Scale Mass Production Solution for Generation 1 & 2 Thin Film Silicon Solar Module. *Proceedings of ISES World Congress 2007 (Vol. I – Vol. V)*, pp. 1220-1223

Turton, R., et al. (2008). *Analysis, Synthesis, and Design of Chemical Processes*. Prentice Hall, ISBN 0-13-512966-4, Westford, MA

Yamamoto, K., et al. (2006). High Efficiency Thin Film Silicon Hybrid Cell and Module with Newly Developed Innovative Interlayer. *Conference Record of the 2006 IEEE 4th World Conference on Photovoltaic Energy Conversion*, pp. 1489-1492

Yang, F., et al. (2009). Uniform growth of a-Si:H / μc-Si:H tandem junction solar cells over 5.7m² substrates. *Processings of 34th IEEE Photovoltaic Specialists Conference*, pp. 1541-1545, Philadelphia, PA

Yang, J., et al. (2005). Amorphous and nanocrystalline silicon-based multi-junction solar cells. *Thin Solid Films* Vol. 487, No. 1-2: pp. 162-169, ISSN 0040-6090

Yang, Y.-T., et al. (2007). The Latest Plasma-Enhanced Chemical-Vapor Deposition Technology for Large-Size Processing. *Journal of Display Technology* Vol. 3, No. 4: pp. 386-391, ISSN 1551-319X

Young, R. (2010). PV Cell Capacity, Shipment and Company Profile Report, IMS Research

6

Analysis of CZTSSe Monograin Layer Solar Cells

Gregor Černivec, Andri Jagomägi and Koen Decock
[1]University of Ljubljana, Faculty of Electrical Engineering,
[2]Department of Materials Science, Tallinn University of Technology,
[3]Solar Cells Department, Ghent University – ELIS,
[1]Slovenia
[2]Estonia
[3]Belgium

1. Introduction

Monograin layer (MGL) solar cell combines the features of a monocrystalline solar cell and a thin film solar cell. The photoactive layer is formed from the kesterite-stannite semiconductor $Cu_2SnZn(S,Se)_4$ (CZTSSe) material with the single-crystalline grains embedded into the epoxy resin (Altosaar et al., 2003). With the graphite back contact, cadmium sulphide (CdS) buffer layer, and zinc oxide (ZnO:Al/i-ZnO) window layer, the remainder of the structure resembles a thin film CIS solar cell in the superstrate configuration (Fig. 1).

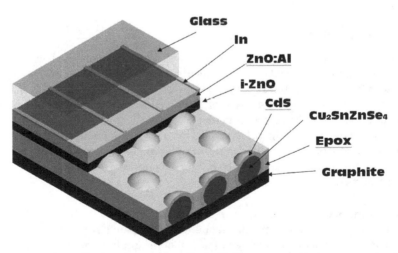

Fig. 1. The MGL solar cell. The photoactive $Cu_2SnZnSe_4$ monograins are covered with CdS, and embedded into the epoxy resin. The thin layer of the intrinsic ZnO serves as the CdS surface passivation and as the barrier for the ZnO:Al impurities. The front contact comprises indium fingers while the back contact is made of the graphite paste.

The main advantage of this cell over the thin film CIGS solar cell are the low production costs – using a relatively simple powder technology (Altosaar et al., 2005), and the replacement of the expensive indium (In) by the less expensive tin (Sn) and zinc (Zn) metals. The photovoltaic properties of this new structure are very promising: the AM1.5 spectrum conversion efficiency reaches up to 5.9% along with the open-circuit voltage (V_{oc}) up to 660 mV and the fill-factor (FF) up to 65%. The short-circuit current (J_{sc}) has its maximal value at the room temperature and then decreases with the lowering temperature. Along with the low FF, these output parameters point to some specific charge transport properties.

In order to discover the origin of the charge transport limiting mechanism we employed the numerical semiconductor simulator Aspin (Topič et al.,1996), based on the drift-diffusion equations (Selberherr, 1984) and coupled to the SRH (Schockley & Read, 1952) recombination statistics. The optical generation rate profile was calculated with the ray tracing simulator Sun*Shine* (Krč et al., 2003), which is able to determine the absorption profile in the illuminated one-dimensional (1D) structure that comprises a stack of layers with flat and/or rough adjacent interfaces. The input semiconductor material parameters were determined from the temperature resolved admittance spectroscopy measurements (Walter et al.,1996): capacitance-voltage (C-V) and capacitance-frequency (C-f), the van der Pauw measurement (Van der Pauw, 1958) and the dark current density-voltage (J-V) characteristics measurements (Sah et al., 1957). The numerical model was implemented in a similar way as in (Černivec et al., 2008) where the measured parameters were used as the input and the J-V and the external quantum efficiency (QE) characteristics were the result of the simulation. By comparing the temperature dependent output characteristics of the AM1.5 illuminated solar cell to the measurements, and additional fine tuning of the input parameters, we assumed the plausible efficiency-limiting mechanism, and by that also revealed the region in the structure that could be responsible for the charge transport limitations.

2. Input parameters measurements

In order to extract material parameters which will be further on used in the numerical analysis, following measurements were conducted: the dark J-V measurement to get insight into the recombination and transport properties of the solar cell, the C-V measurement which indicates the width and the shape of the junction, and the C-f measurement which results the information of the defect properties of the semiconductor material. The common assumption in the analyses of the measurements is a single-junction model of the solar cell. In the interpretation of the Van der Pauw measurement results we assumed a similar morphology of the annealed tablet of the CZTSSe material as it is one in the solar cell's monograin absorber.

2.1 One-diode model
Calibration of the parameters of the one-diode model does not yield any input parameters for our numerical model, but it rather gives us initial insight into the transport properties of the MGL solar cell. Table I contains the extracted temperature dependent parameters of the fitted one-diode model (Sze & Ng, 2007). The high ideality factors (n_{id}) of the temperature dependent dark J-V measurement indicate the CdS/CZTSSe heterointerfacial limited transport.

T [K]	J_0 [mA/cm^2]	n_{id} [/]	R_s [Ωcm^2]	G_{sh} [mS/cm^2]
310	1.08x10^{-3}	2.68	2.10	0.23
290	4.85x10^{-4}	2.78	2.36	0.17
270	2.50x10^{-4}	2.99	2.66	0.12
250	8.93x10^{-5}	3.19	3.44	0.083
230	3.30x10^{-5}	3.37	4.37	0.061
210	1.44x10^{-5}	3.78	6.65	0.033

Table 1. Parameters of the fitted one-diode model.

The ideality factors above 2 deviate from the standard Sah-Noyce-Shockley theory (Sah et al., 1957) and point either to the tunnelling enhanced recombination in the space charge region (SCR) (Dumin & Pearson, 1965) or to the multilevel recombination (Breitenstein et al., 2006; Schenk et al., 1995) occurring in the highly defective interfacial regions.

Fig. 2 shows the Arrhenius plot of the dark saturation current (J_0) and its extracted activation energy ($E_{A, J0}$). The activation energy is the distance between the Fermi level and the edge of the minority carrier energy band, since these are responsible for the recombination current. In the case of the MGL solar cell, at the CdS/CZTSSe heterointerface the inverted surface makes holes to be the minority carriers, Fig. 8. Thus the $E_{A, J0}$ represents the energy distance between the CZTSSe absorber's valence band and the Fermi level near the heterointerface.

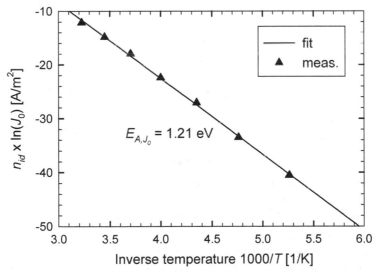

Fig. 2. Arrhenius plot of the dark saturation current as obtained from the one-diode model. The slope of the ideality factor weighted logarithm of the dark saturation current versus the inverse absolute temperature, results the activation energy $E_{A, J0}$. T is temperature in Kelvin.

Comparing the value of the $E_{A, J0}$ (Fig. 2) to the absorber's band-gap energy as extracted from the QE measurement (Fig. 13, $E_{g,CZTSSe}$ = 1.49 eV), this indicates the position of the recombination peak near the heterointerface and inside the SCR – as depicted in Fig. 8.

2.2 Capacitance-voltage measurement

To obtain the approximate values of the concentration of the uncompensated acceptors (Kosyachenko, 2010) at the edge of the SCR, and the hole mobility ($\mu_{h,CZTSSe}$) of the CZTSSe absorber layer, we combined the temperature resolved C-V and the van der Pauw measurements. Since the concentration of the uncompensated acceptors at the edge of the SCR corresponds to the density of free holes, we will further on introduce this as new parameter called the "apparent doping" – p_{SCR}.

Fig. 3 shows the temperature and the bias voltage dependent capacitance plot – the Mott-Schottky plot, where the capacitance results from the admittance measurement at 10 kHz. The nonlinear curves in the Mott-Schottky plot indicate a spatially non-uniform p_{SCR}, while their temperature trend points to the temperature decreasing capacitance. The slope of the curves at $V = 0$ V indicates that, in dark conditions, the apparent doping at the edge of the SCR gradually increases with the decreasing temperature.

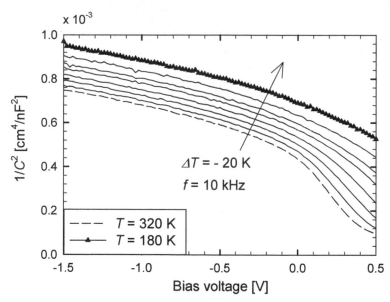

Fig. 3. The Mott-Schottky plot at 10 kHz. The dashed curve correlates to the temperature at 320 K. Arrow indicates the trend of the temperature decrement. The temperature step equals to 20 K. All curves are measured with a small signal of 10 kHz.

When we observe the 0V bias points as depicted in Fig. 4 by the triangles, we can see that p_{SCR} decreases when moving from the quasi-neutral region towards the SCR. However, for the higher temperatures (320 K, 300 K) p_{SCR} seems to be increasing towards the heterointerface after it has reached its minimum value. We are not able to explain this trend properly, but since the increasing p_{SCR} towards the heterointerface would produce only a poor photovoltaic junction, in the modelling we use the p_{SCR} values as obtained at 0 V bias. The trend of the increasing SCR width along with the increasing p_{SCR} could results from the influences of the non-ideally asymmetrical n^+/p (CdS/CZTSSe) junction in which the SCR extends also into the n^+ buffer region (CdS).

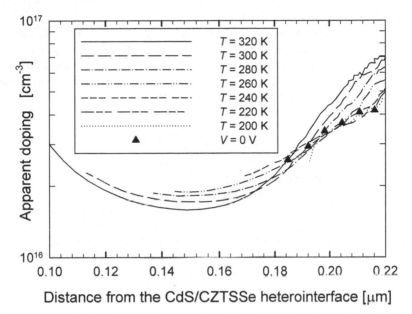

Fig. 4. The apparent doping density p_{SCR} obtained from the bias voltage derivative of the Mott-Schottky plot. The distance from the junction is calculated from the space charge region capacitance. Triangles depict the 0 V bias conditions.

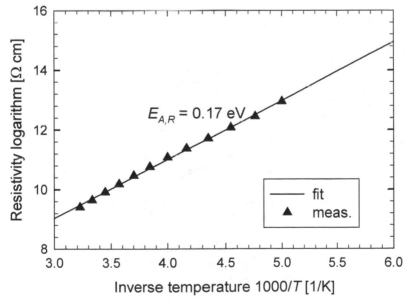

Fig. 5. Arrhenius plot of the van der Pauw measurement conducted on the annealed CZTSSe tablet. $E_{A,R}$ is the extracted activation energy. T is temperature in Kelvin.

2.3 Van der Pauw measurement

The van der Pauw measurements were conducted on the tablet of the annealed CZTSSe monograin material. The Arrhenius plot of the resistivity (ρ) of the monograin material tablet (Fig. 5) reveals the thermal activation energy ($E_{A,R}$) equal to 0.17 eV, and a very low hole mobility $\mu_{h,CZTSSe}$ equal to 0.02 cm²/Vs at 310 K. The latter was calculated according to (1) and using the p_{SCR} as obtained from the C-V profiling:

$$\mu_{h.CZTSSe} = \frac{1}{q \cdot p_{SCR} \cdot \rho}. \tag{1}$$

2.4 Capacitance-frequency measurement

Plotting the capacitance as a function of the measurement frequency on a semi-logarithmic scale can reveal some defects present in the energy gap of the CZTSSe absorber layer of the MGL solar cell. A gradually decaying capacitance indicates a defect with a broad energy band, while a steep transition indicates a single level defect (Burgelman & Nollet, 2005). The temperature resolved C-f plot shown in Fig. 6 reveals both types of transitions: a gradually decreasing capacitance at the high temperature limit (indicated with triangles), and a characteristic inflection point at the frequency equal to 10 kHz in the low temperature limit (indicated with circles).

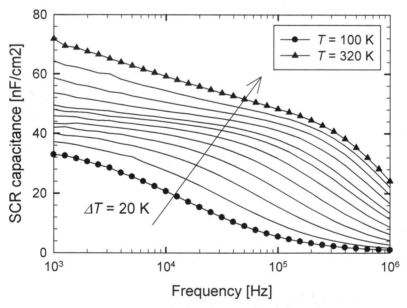

Fig. 6. Frequency dependent space charge region's capacitance measured at 0.2 V of forward bias. Solid curve with circles depicts the relation at 100 K. The arrow indicates the trend of the curves with the increasing temperature. The temperature step equals to 20 K. The curves at lower temperatures exhibit pronounced inflection points thus indicate emission from shallow traps.

The decreasing capacitance going from the high temperature towards the low temperature indicates the 'freeze out' of the carriers located in the deep traps: the temperature shrinking of the Fermi distribution tail makes the deep trapped charge less sensitive to the small perturbations of the Fermi level (the applied ac signal). The analysis according to (Walter et al.,1996) reveals two trap distributions which are shown in Fig. 7. Measurement at room temperature senses a broad trap distribution extending at least 0.3 eV deep into the energy gap from the valence band, while the measurement at low temperature fingers a very narrow distribution with its maximum at 0.05 eV. Since this maximum remains present also at high reverse biases (not shown here), we believe that this trap extends throughout the whole CZTSSe absorber layer and acts as the intrinsic acceptor doping level. However we can not draw any strong conclusions on the type of the deep trap distribution, but since this could be responsible for the compensating effect; we postulated it to be the donor-like.

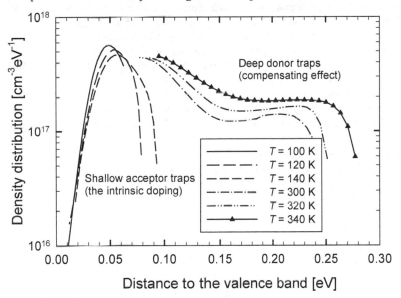

Fig. 7. Trap density distributions extracted at 0.2 V of forward bias calculated as the frequency derivative of the space charge region's capacitance. Calibration parameters were chosen according to (Walter et al.,1996): $U_d = 0.8$ V (built-in voltage), $\beta_p N_V = 5\times10^7$ Hz (trap emission coefficient), $E_{fp} = 0.7$ eV (the Fermi level position relating the valence band), $1\times10^3 \leq f \leq 1\times10^6$ Hz (frequency range).

In Fig. 7 the pronounced narrow distribution at 0.05 eV above the valence band indicates the shallow acceptor traps responsible for the intrinsic doping, while the deeper and wider donor distribution (marked with triangles) results the compensation effect.

3. Modelling

From the measurements we obtained a certain insight into the recombination and transport properties (the dark J-V and the Van der Pauw measurements), the doping profile (C-V measurement) and the indication of the shallow traps (C-f measurement). These will be used as the guidelines to define the numerical model of the CZTSSe MGL solar cell.

The 1D carrier transport model can accurately describe the current flow only in the direction vertical to the layered structure (the direction orthogonal to the solar cell plane) therefore following assumptions are made: i) current flow in the matrix plane between the adjacent monograins is neglected, ii) all the semiconductor parameters are meant as the "effective parameters", thus neglecting the morphology by transforming a single spherical monograin solar cell into the 1D rod, and iii) the "spatial fill-factor" (S_{FF}) is introduced, which is the ratio of the grain covered area to the whole contact area. It is important to note that the S_{FF} affects only the extensive solar cell parameters (J_{sc}) while the intensive parameters (V_{oc}, FF and QE) remain intact. In our case the S_{FF} equals to 0.78.

The most important semiconductor parameters which have to be defined for each layer of the MGL solar cell prior to simulation are the band-gap energy (E_g), the electron affinity (E_χ), the acceptor and/or donor doping (N_A, N_D), the hole and electron low-field mobility (μ_h, μ_e), the hole and electron effective masses (m_h, m_e), and the parameters of the traps and/or the recombination centres (N_t – distribution density, E_t – distance to the valence band, σ – trap cross section, e_t – characteristic energy). By analyzing the conducted measurements (C-V, van der Pauw, C-f) we extracted the initial values of these parameters, relating to the CZTSSe absorber and/or to the CdS/CZTSSe heterointerface. These were further on subjected to the calibration procedure in order to fit the dark structure and the illuminated structure output characteristics to the measurements (J-V and QE). The rest of the absorber and heterointerface parameters, and those relating to the window (ZnO:Al/ZnO) and buffer (CdS) layers of the MGL solar cell, were taken similar to those used in (Černivec et al., 2008).

3.1 Dark structure *J-V* characteristics

Fig. 8 shows the CZTSSe MGL solar cell structure in its thermodynamic equilibrium. The complete solar cell comprises glass(2 mm)/ZnO:Al(1.6 µm)/i-ZnO(200 nm)/CdS(50 nm)/CZTSSe(60 µm)/graphite(500 nm) layers with the additional 100 nm thick surface

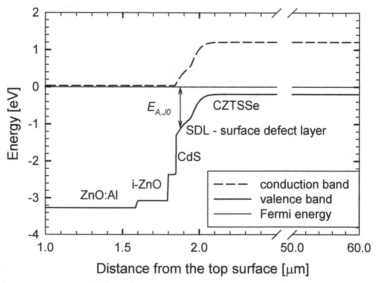

Fig. 8. Energy band diagram of CZTSSe solar cell in thermodynamic equilibrium at 310 K. $E_{A,J0}$ indicates the recombination activation energy as obtained from the Arrhenius plot from Fig. 2.

defect layer (SDL) between the CdS and the CZTSSe to account for the interfacial defects. Because of the degenerate position of the Fermi level in the ZnO:Al, i-ZnO and CdS layers, we assume these will act as the emitter contact, while the graphite at the back acts as the ohmic base contact. Further on in the structure we introduce the SDL which has an increased concentration of the mid-gap defects of the donor ($N_{tD,SDL}$) and the acceptor ($N_{tA,SDL}$) types. $N_{tD,SDL}$ will be responsible for the recombination current while the $N_{tA,SDL}$ will set the Fermi level position in the SDL layer and thus activate the $N_{tD,SDL}$.

The van der Pauw measurements of the sole CZTSSe tablets exhibit unusual high resistances, thus we assume that $\mu_{h,CZTSSe}$ will have an important impact to the series resistance – R_s (Table I). The Arrhenius plot in Fig. 5 shows the latter's exponential dependence on temperature, revealing the activation energy of 0.17 eV. We believe that the high R_s originates from the compensation of the shallow acceptor doping ($N_{tA,CZTSSe}$) by the broader distribution of deeper donor levels ($N_{tD,CZTSSe}$). This agrees well with the C-f measurement results shown in Fig. 7. Therefore, rather than calculating the mobility from the van der Pauw measurement, we will use a numerical fitting procedure to calibrate the $\mu_{h,CZTSSe}$ and the $N_{tD,SDL}$ for the preselected values of the $N_{tA,CZTSSe}$ and the $N_{tD,CZTSSe}$. The initial values for the latter two were calculated from the C-f measurement (Fig. 7).

Fig. 9 shows the calibration procedure of the measured and the simulated dark J-V characteristics at 310 K. By increasing the total concentration of the SDL mid-gap donor defects ($N_{tD,SDL}$) the dark saturation current increases, as shows the inset of Fig. 9. In the voltage range from 0.4 V to 0.6 V a good J-V fit can be found for the $N_{tD,SDL}$ equal to 10^{18} cm^{-3}, but still expressing a deviation in the slope as the result of the non-matching ideality factors: with this model it is not possible to obtain such a high ideality factor as yielded the measurement-calibration in Table I. For the lower applied voltages ($V < 0.4$ V) there is a significant deviation in characteristics which can be attributed to the shunt conductance. To compensate this difference the external shunting element can be added in the model, using the value equal to the G_{sh} at 310 K (Table I). A very good fit is found in the voltage range $V > 0.5$ V by setting the value of the $\mu_{h,CZTSSe}$ to 1.5 cm^2/Vs – indicated by the solid line in Fig. 9.

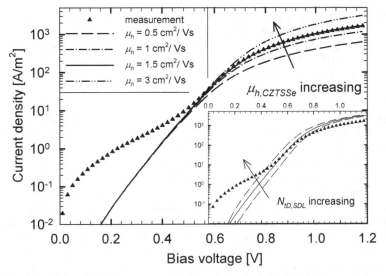

Fig. 9. Calibration of the CZTSSe monograin layer's and of the SDL's transport parameters.

In Fig. 9 the calibrated value of the CZTSSe hole mobility equals to 1.5 cm^2(Vs)$^{-1}$ and the corresponding electron mobility equals to 8 cm^2(Vs)$^{-1}$. The inset of Fig 9 shows calibration of the SDL defect concentration. The calibrated defect concentration ($N_{tD,SDL}$ = 8x10^{19} cm^{-3}/eV) corresponds to the solid J-V curve of the three simulated characteristics. The J-V curve above (dash-dotted) and the J-V curve below (dashed) correspond to one order of magnitude higher and to one order of magnitude lower SDL defect concentration, respectively.

To summarize the dark model, this is valid for the bias voltages higher than 0.5 V. When the solar cell is illuminated, this usually happens to be the range at which the recombination current starts to compensate the photogenerated current, and therefore important to match the correct V_{oc} value. For the bias voltages lower than 0.5 V the recombination current is rather low and the photogenerated current will dominate the J-V characteristics. Thus the external G_{sh} might be of lesser importance when observing the illuminated solar cell structure.

3.2 Illuminated structure characteristics

In order to calibrate the CZTSSe solar cell model under illumination, we choose to observe the temperature behaviour of the J_{sc}. This is mainly determined by the collection efficiency of the photogenerated carriers in the SCR. The collection efficiency in a large extend depends on the width of the SCR (Fig. 4), determined by the shallow acceptor traps in the CZTSSe - $N_{tA,CZTSSe}$, while its temperature dependence governs the occupation function F of the deeper donor traps $N_{tD,CZTSSe}$ (Fig. 10). Fig. 10 shows the $N_{tA,CZTSSe}$ and $N_{tD,CZTSSe}$ distributions similar to the measured trap densities from Fig. 7, and the occupation function F at 310 K and 210 K. The peak values of the trap distributions are not the same as the measured traps, but were rather subjected to the calibration procedure of fitting the J-V and QE measured and simulated characteristics. At the edge of the SCR the apparent doping p_{SCR} is a result of the compensatory effect of the density of the occupied $N_{tA,CZTSSe}$ and the density of the unoccupied $N_{tD,CZTSSe}$:

$$p_{SCR} = N_{tA,CSZSSe} - (1 - F) \cdot N_{tD,CSZSSe} . \tag{2}$$

When temperature decreases the E_{fp} moves towards the valence band, what creates more deep donors unoccupied (f_B decreases), and lowers the p_{SCR}.

In Fig. 10 the trap distributions of the model are calibrated to fit the measured short-circuit current density at 310 K. The distributions correlate well with the calculated distributions shown in Fig. 7. On the right axis the occupation functions at two different temperatures are shown in order to explain the temperature dependent collection efficiency and its influence to the short-circuit current.

The temperature decreasing p_{SCR} decreases the SCR width, leading into the lower collection efficiency and lower J_{sc}. Fig. 11 shows the SCR narrowing as the result of the Fermi redistribution according to Fig. 10. The decreased p_{SCR} would normally lead into the wider SCR, if the net charge of the SDL remained constant. This would be the case with the ideal asymmetrical n^+/p junction, resulting from the shallow doping levels. But since the net charge in the SDL originates also from the deep defects, these are then affected by the change of the charge in the CZTSSe layer. Therefore in order to satisfy the Poisson's balance, the lower temperature also leads into the charge redistribution in the SDL layer (omitted for clarity in Fig. 11): the decrement of the negative charge resulting from the less occupied acceptor traps in the CZTSSe layer is balanced by the decrement of the positive charge from the deep defects in the SDL. In the SDL the temperature shift of the Fermi level towards the conduction band makes the deep donor defects less ionized and increases the ionization of the deep acceptors.

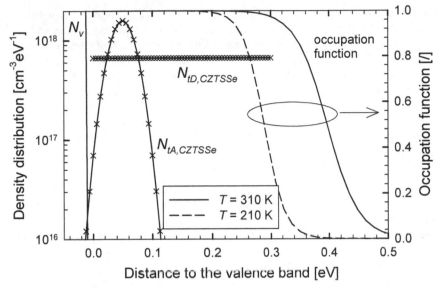

Fig. 10. Trap distributions of the CZTSSe monograin layer 50 nm deep in the SCR from the SDL/CZTSSe heterointerface.

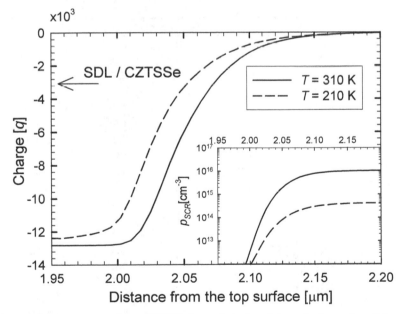

Fig. 11. Space charge region of the CZTSSe layer (q is the electron's charge) and its temperature dependence resulting from the occupation function variation (shown in Fig. 10). On the left, the interface to the SDL is indicated. The inset shows the temperature variation of the apparent doping p_{SCR}.

The modelled SCR width of approximately 0.2 μm and the p_{SCR} concentration of 10^{16} cm^{-3} at 310 K agree well with the respective measured values which equal to 0.18 μm and 2.6x10^{16} cm^{-3}, as observed from Fig. 4. Similarly well agrees the temperature correlation between the p_{SCR} and the SCR width: with the increasing p_{SCR} also the increasing SCR width is observed. In the measurement this correlation is indicated with the triangles (Fig. 4). However the corresponding temperatures do not comply: in the measurement the 320 K triangle corresponds to the lowest p_{SCR} and the 220 K triangle corresponds to the highest p_{SCR}. One should indeed always take care about the width of the SCR calculated in the apparent doping density analysis. There, the following formula is used to calculate the SCR width:

$$W_{SCR} = \frac{\varepsilon}{C}, \tag{3}$$

where ε is the permittivity and C is the capacitance. This formula however only holds if the capacitance is governed by the depletion, and not by filling and emptying of deep states. As can be seen in Fig. 6 the capacitance is indeed governed by defects rather than depletion at f=10kHz.

Table 2 summarizes the calibrated material parameters. The parameters which were the subject of calibration are denoted bold, while the dash corresponds to the parameter for which we used the value 0. In the reality this would correspond to a very low value. Other material parameters are similar as in (Černivec et al., 2008). The effective density of states is calculated from the corresponding effective masses (Sze & Ng, 2007).

	ZnO:Al	i-ZnO	CdS	SDL	CZTSSe
W [μm]	1.6	0.2	0.05	0.1	60
m_e [m_0]	0.27	0.27	0.27	0.09	0.09
m_h [m_0]	0.78	0.78	0.78	0.73	0.73
N_D [cm^{-3}]	10^{18}	10^{18}	10^{18}	-	-
N_A [cm^{-3}]	-	-	-	-	-
E_g [eV]	3.3	3.1	2.4	1.4	1.4
E_χ [eV]	4.0	4.0	4.0	4.0	4.0
E [ε_0]	9.0	9.0	9.0	13.6	13.6
μ_e [cm^2/Vs]	100	100	100	**40**	**8**
μ_h [cm^2/Vs]	25	25	25	**15**	**1.5**
N_{tA} [cm^{-3}/eV]	-	-	-	**4x10^{17}**	**8x10^{16}**
E_{tA} [eV]	-	-	-	**0.7**	**0.05**
σ_{nA} [cm^2]	-	-	-	**2x10^{-15}**	**2x10^{-15}**
σ_{pA} [cm^2]	-	-	-	**8x10^{-13}**	**8x10^{-13}**
e_{tA} [eV]	-	-	-	**0.1 (step)**	**0.02 (gauss)**
N_{tD} [cm^{-3}/eV]	-	-	-	**8x10^{19}**	**7x10^{17}**
E_{tD} [eV]	-	-	-	**0.7**	**0.15**
σ_{nD} [cm^2]	-	-	-	**8x10^{-14}**	**10^{-14}**
σ_{pD} [cm^2]	-	-	-	**2x10^{-15}**	**10^{-15}**
e_{tD} [eV]	-	-	-	**0.1 (step)**	**0.3 (step)**

Table 2. Material parameters of the CZTSSe MGL monograin layer solar cell.

4. Analysis of the model

Fig. 12 shows the measured and simulated J-V characteristics of the CZTSSe MGL solar cell. The measured characteristics were obtained from the I-V characteristics normalized to the contacting area of the solar cell equal to $A = 4.81$ mm^2. Here we used the assumptions i) and ii) as defined in 3. Since the CZTSSe monograins shape in the spherical forms this means that the real current density varies throughout the structure. In the simulated J-V characteristics at 310 K we also took into account the S_{FF} as the assumption iii). This means that the J_{sc} obtained by using the parameters from Table 2 would in fact be larger by this factor.

In Fig. 12 we can observe a very good agreement of the measured and simulated J-V characteristics at 310 K while the simulation at 210 K exhibits a discrepancy in all solar cell output parameters. A possible reason for the non-matched J_{sc} at 210 K could be that in the modelling we did not account for the temperature dependent mobility, which could be the case as seen from the van der Pauw measurement of the monograin tablet (Fig. 5).

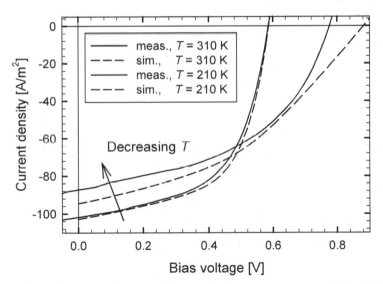

Fig. 12. Comparison of the measured and simulated J-V characteristics of the AM1.5 illuminated CZTSSe monograin solar cell.

Dashed lines in Fig. 12 represent the simulation and the arrow indicates temperature decrement. The short-circuit current and the open-circuit voltage trends are well correlated while their absolute value deviation at the low temperature indicate the necessity to include the temperature dependent mobility and the tunnelling enhanced recombination, respectively. At 210 K a significant mismatch also occurs with the V_{oc}. This leads us to the conclusion that it is not merely the SRH recombination (Sze & Ng, 2007) in the SDL layer that limits the V_{oc}, but there should also be present other recombination mechanisms which are less thermodynamically affected, namely the tunnelling enhanced recombination (Dumin & Pearson, 1965). The tunnelling enhanced recombination would reduce the rate of the V_{oc}-T change.

The optical simulations were performed using the Sun*Shine* simulator (Krč et al., 2003) which takes as an input a layered structure with the wavelength dependent complex refraction index coefficients, which comprise the real part $n(\lambda)$, called refractive index, and the complex part $k(\lambda)$ known as the extinction coefficient. Both are defined in for each layer. For the monograin material we used the complex refraction index coefficients as obtained by Paulson (Paulson et al., 2003) for the thin film $Cu(In_{1-x}Ga_x)Se_2$ alloy with the $x = 0.66$. This corresponds to the energy gap of 1.41 eV. The layer's interfaces were described using the roughness coefficient – σ_{rms}. In our case we set the σ_{rms} equal to 100 nm at all interfaces. Simulation of the external quantum efficiency (Fig. 13) shows a good agreement between the measured QE and the simulated QE in the shorter wavelengths region, while in the middle wavelengths there seems to exist some discrepancy – most probably due to the discrepancy between the measured and modelled $\mu_{h,CZTSSe}$. The cut-off wavelengths are well pronounced at both temperatures and correspond to the band-gap of 1.4 eV. In the long wavelength region ($\lambda > 900$ nm) the non-vanishing plateau of the simulated QE points to a mismatch in the absorption properties of the thin film CIGS and the monograin layer CZTSSe materials.

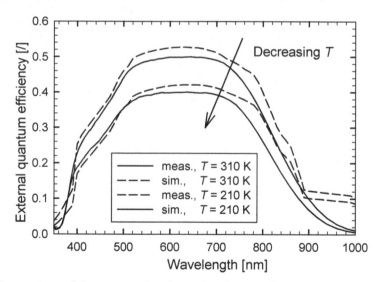

Fig. 13. Comparison of the measured and simulated external quantum efficiency of the AM1.5 illuminated CZTSSe monograin solar cell.

In Fig. 13 dashed lines represent the simulation and the arrow indicates temperature decrement. The non-vanishing plateau of the simulation originates from the mismatch in the absorption properties of the thin film CIGS (used in the simulation) and the monograin layer CZTSSe materials.

Both, measured and simulated QE show that the temperature change does not affect their shape, which inclines us to a conclusion that most of the photogenerated carriers recombine in the SDL and at the SDL/CZTSSe interface. This fact can as well be observed from the cumulative recombination profile (not shown here).

The absorptance simulations show that if all photogenerated carriers originating from the photon flux absorbed in the CZTSSe layer were extracted, the J_{sc} would equal to 37.7 mA/cm². Taking into account the S_{FF} the latter would reduce to a 29.4 mA/cm². This value is still about 3 times larger than the measured (simulated) J_{sc} at 310 K, showing tremendous possibilities in improvement of the collection efficiency of the monograin CZTSSe absorber.

5. Conclusion

We have set up the baseline model of the $Cu_2SnZn(Se,S)_4$ monograin layer solar cell, which is able to predict the J-V characteristics and the external QE of the AM1.5 illuminated MGL solar cell in the temperature range from 310 K to 210 K. The model comprises following material properties:
i) in between the CdS and CZTSSe layers, the highly defective region called the surface defect layer – SDL, comprising a high concentrations of the mid-gap donor defects and a lower concentration of the mid-gap acceptor defects;
ii) in the CZTSSe monograin layer the narrow Gaussian distribution of shallow acceptor traps at 0.05 eV above the valence band and the wider distribution of the compensatory donor traps extending at least 0.3 eV deep into the energy band, relative to the valence band;
iii) energy gap of the CZTSSe monograin material equals to 1.4 eV, width of the SCR at 310 K equals to 180–200 nm and the concentration of the apparent doping p_{SCR} is in the range from 1×10^{16} cm^{-3} to 2×10^{16} cm^{-3}.
Low FF can be attributed to the low CZTSSe hole mobility, which equals to 1.5 cm^2/Vs, and to the low apparent doping p_{SCR}, which originates from the compensatory effect of the shallow acceptors and deeper donors. Comparison of the flux absorbed in the CZTSSe monograin absorber and the three times lower actual current density of the extracted carriers shows us that further possibilities may reside in the shaping of the collection efficiency of the monograin absorber and/or in the additional passivation of the CdS/CZTSSe interface. Since the former is mainly attributed to the SCR this might not be an easy technological task. Whether these limiting properties are the result of the necessary surface engineering prior to the formation of the CdS/CZTSSe monograin heterojunction or they simply originate from the physical properties of the structure's materials, we were be not able to determine at this point.

6. Acknowledgments

Authors would like to thank prof. dr. Jüri Krustok, Tallinn University of Technology, for his objective criticism which helped to improve the quality of this work. We also thank prof. dr. Marko Topič, University of Ljubljana, for his approval on the use of the simulation software Aspin2 and SunShine.

7. References

Altosaar, M.; Jagomägi, A.; Kauk, M.; Krunks, M.; Krustok, J.; Mellikov, E.; Raudoja, A. & Varema, T. (2003). Monograin layer solar cells. *Thin Solid Films*, Vol. 431-432, pp. 466-469, ISSN 0040.6090
Altosaar, M.; Danilson, M.; Kauk, M.; Krustok, J.; Mellikov, E.; Raudoja, J.; Timmo, K. & Varema, T. (2005). Further development in CIS monograin layer solar cells technology. *Solar Energy Materials & Solar Cells*, Vol. 87, pp. 25-32, ISSN 0927.0248
Breitenstein, O.; Altermatt, P.; Ramspeck, K. & Schenk, A. (2006). The origin of ideality factors N>2 of shunt and surfaces in the dark I-V curves of SI solar cells, *Proceedings*

of the 21st European Photovoltaic Solar Energy Conference, pp. 625-628, Dresden, Germany.

Burgelman, M. & Nollet, P. (2005). Admittance spectroscopy of thin film solar cells. *Solid State Ionics*, Vol. 176, pp. 2171-2175, ISSN 0167.2738

Černivec, G.; Jagomägi, A.; Smole, F. & Topič, M. (2008). Numerical and experimental indication of thermally activated tunnelling transport in CIS monograin layer solar cells. *Solid State Electronics*, Vol. 52, pp. 78-85, ISSN 0038.1101

Dumin, D.J. & Pearson, G.L. (1965). Properties of gallium arsenide diodes between 4.2 and 300 K. *Journal of Applied Physics*, Vol. 36, No. 11, pp. 3418-3426, ISSN 0021.8979

Kosyachenko, L. (2010). Efficiency of thin film CdS/CdTe Solar Cells, In: *Solar Energy*, R.D. Rugescu, (Ed.), 105-130, InTech, ISBN 978-953-307-052-0, Vukovar, Croatia.

Krč, J.; Smole, F. & Topič M. (2003). Analysis of light scattering in amorphous Si:H solar cells by one-dimensional semi-coherent optical model. *Progress in Photovoltaics: Research and Applications*, Vol. 11, pp. 15-26, ISSN 1062.7995

Paulson, P.D.; Birkmire, R.W. & Shafarman, W.N. (2003). Optical characterization of $CuIn_{1-x}Ga_xSe_2$ alloy thin films by spectroscopic ellipsometry. *Journal of Applied Physics*, Vol. 94, No. 2, pp. 879-888, ISSN 0021.8979

Sah, C.T.; Noyce, R.N. & Shockley, W. (1957). Carrier generation and recombination in p-n junctions and p-n junction characteristics. *Proceedings of the Institute of Radio Engineers*, Vol. 45, No. 9, pp. 1228-1243, ISSN 0731.5996

Schenk, A. & Krumbein, U. (1995). Coupled defect-level recombination: Theory and application to anomalous diode characteristics. *Journal of Applied Physics*, Vol. 78, No. 5, pp. 3185-3192, ISSN 0021.8979

Schockley, W.; Read, W.T. (1952). Statistics of the recombination of holes and electrons. *Physical Review*, Vol. 87, pp. 835-842

Selberherr, S. (1984). *Analysis and simulation of semiconductor devices*, Springer Verlag, ISBN 978.0387818009, Vienna, Austria

Sze, S.M. and Ng, K.K. (2007). *Physics of semiconductor devices*, John Wiley & Sons, ISBN 9971.51.266.1 , New Jersey, USA

Topič, M.; Smole, F. & Furlan, J. (1996). Band-gap engineering in $CdS/Cu(In,Ga)Se_2$ solar cells. *Journal of Applied Physics*, Vol. 79, No. 11, pp. 8537-8540, ISSN 0021.8979

Van der Pauw, L.J. (1958). A method of measuring specific resistivity and Hall effect of discs of arbitrary shape. *Phillips Research Reports*, Vol. 13, pp. 1-9, ISSN 0031.7918

Walter, T.; Herberholz, R.; Müller, C. & Schock H.W. (1996). Determination of defect distributions from admittance measurements and application to $Cu(In,Ga)Se_2$ based heterojunctions. *Journal of Applied Physics*, Vol. 80, No. 8, pp. 4411-4420, ISSN 0021.8979

7

Novel Deposition Technique for Fast Growth of Hydrogenated Microcrystalline Silicon Thin-Film for Thin-Film Silicon Solar Cells

Jhantu Kumar Saha[1,2] and Hajime Shirai[1]

[1]Department of Functional Material Science & Engineering,
Faculty of Engineering,
Saitama University,
[2]Current address: Advanced Photovoltaics and Devices,(APD) Group,
Edward S. Rogers Sr. Department of Electrical and
Computer Engineering, University of Toronto,
[1]Japan
[2]Canada

1. Introduction

The microcrystalline silicon material is reported to be a quite complex material consisting of an amorphous matrix with embedded crystallites plus grain boundaries. Although this material has a complex microstructure, its optical properties have a marked crystalline characteristic: an optical gap at 1.12 eV like c-Si. This implies the spectral absorption of µc-Si:H covers a much larger range than a-Si:H which posses an optical gap between 1.6 and 1.75eV[i]. Compared to a-Si:H that absorbs light up to 800 nm, µc-Si:H absorbs light coming from a wider spectral range, extending up to 1100 nm . On the other hand, within its range of absorption, the absorption of a-Si:H is higher than that of µc-Si:H –due to the indirect gap of the latter. Therefore, the optical combination of these two materials takes advantage of a larger part of the solar spectrum (compared to a single-junction cell) and the conversion efficiency of the incident light into electricity can be consequently improved. Furthermore, the µc-Si:H solar cell is reported to be largely stable against light induced degradation and enhanced carrier mobility in contrast to amorphous silicon films counterpart. Consequently hydrogenated microcrystalline silicon is one of the promising materials for application to thin-film silicon solar cells.

2. Growth techniques of hydrogenated microcrystalline silicon

The growth of µc-Si:H material uses silane (SiH_4) and hydrogen as source-gas. It is currently admitted that free radical precursors (SiH_x)-SiH_3 is suspected to favor the µc-Si:H growth-and H-enhances crystalline growth by etching of looser a-Si:H tissue-were needed to attain microcrystalline growth. In order to obtain such reactive species, decomposition of the source-gases is necessary. At first, this was obtained by using PE-CVD at high temperatures (600°C). The use of low deposition temperatures of 200-300°C with a plasma present in the deposition chamber, the so called Plasma-Enhanced Chemical Vapor Deposition technique (PE-CVD) was developed later on and allowed the low-temperature deposition of μc-Si:H films, and rapid progresses have been achieved. Unfortunately, "state-of-the-art" microcrystalline silicon solar cells consist of intrinsic μc-Si:H layers that are deposited by rf and VHF PE-CVD at deposition rates of only 1-5 Å/s. On the other hand, a μc-Si:H film with a 2-μm- thickness intrinsic absorption layer is required for application to Si thin-film solar cells because of the low optical absorption in the visible region. The μc-Si:H i-layer deposition step is the most time consuming step in the deposition sequence of the solar cell. Therefore, a novel fast deposition technique of μc-Si:H is required.

3. Novel fast deposition techniques of microcrystalline silicon

Now-a- days, for the high throughput of high-efficiency μc-Si solar cells in PV industry, one of the most crucial requirements is fast deposition of μc-Si without deteriorating the optical, structural and electronic properties of the film. To overcome the difficulty, several high-density plasma sources have been developed, such as very high frequency (VHF) plasma, inductive coupling plasma (ICP) and surface wave plasma (SWP). As it has been reported, the excitation frequency of a plasma source has an important effect on the electron acceleration in the plasma, and a high excitation frequency is expected to result in a high electron density and a low electron temperature. Therefore two new microwave plasma sources have been developed i.e. Low-pressure high-density microwave plasma source utilizing the spoke antenna and the remote-type high-pressure microwave plasma using a quartz tube having an inner diameter of 10 mm and applied those for the fast deposition of μc-Si films for Si thin-film solar cells. The remote-type high-pressure microwave plasma will be discussed in elsewhere.

3.1 Low-pressure high-density microwave plasma source utilizing the spoke antenna
The microwave plasma source is shown schematically in Fig. 1, which is composed of the combination of a conventional microwave discharge and a spoke antenna. Its chamber size is 22 cm in diameter, which enables large-scale film processing. The spoke antenna is located on a 15 mm-thick quartz plate, which is not inside of the vacuum chamber. The antenna system is shown in Fig. 2 more in detail. The length of each spoke is 4 cm, which is about 1/4 of the wavelength of a 2.45 GHz wave. The design of the spoke antenna assembly

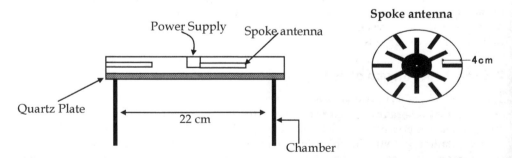

Fig. 1. A schematic illustration of the microwave plasma source

Novel Deposition Technique for Fast Growth of Hydrogenated Microcrystalline Silicon Thin-Film
for Thin-Film Silicon Solar Cells

131

is based on an inter-digital filter composed of parallel cylindrical rods (spokes) arranged between parallel-grounded plates. The spokes are resonantly coupled by the stray capacitance between adjacent spokes and the inductance of the spokes themselves. The resonance condition of an introduced angular frequency is given by $\omega=2\pi f=1/(C \times L)^{1/2}$, where f is the introduced frequency, C is the array capacitance, and L is the antenna inductance. Thus, the antenna operates as a band-pass filter. The spokes are arranged like those in a wheel, and the plasma serves as one of the grounded plates. The electromagnetic wave propagates through the spokes consecutively with a phase difference of 90°, and microwave current flows in every spoke. The current in the spokes couples inductively and capacitively to the plasma ("CM coupling"), and the induction current in the plasma accelerates the electrons to sustain the plasma, as shown in Fig. 2 & 3. The power is supplied from the center of the antenna, and the plasma under the spoke antenna is radially discharged because induction current flows near every spoke. As a result, uniform microwave plasma over an area of diameter greater than 20 cm can be generated efficiently. As well, since no magnetic field is required to generate the high-density microwave plasma, it is possible to design a simple source yielding high-density and low-temperature plasma.

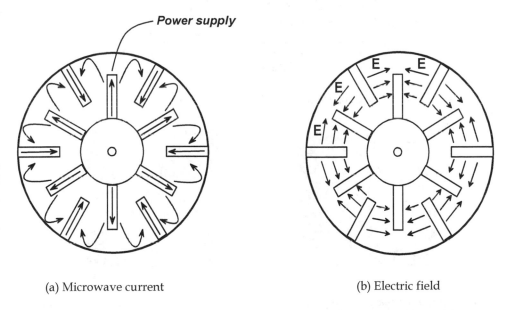

(a) Microwave current   (b) Electric field

Fig. 2. The newly developed spoke antenna for introduction of microwave power (a) Microwave current, (b) Electric field.

From a material processing standpoint, large-area microwave plasmas (MWPs) have several advantages in comparison with other types of high-density sources. First, MWPs, being no magnetized sources, are free from such magnetic field induced problems as inhomogeneous density profile and charge-up damage, which is often, experienced in electron cyclotron resonance (ECR) or helicon plasma sources. Second, MWPs can be enlarged to diameters

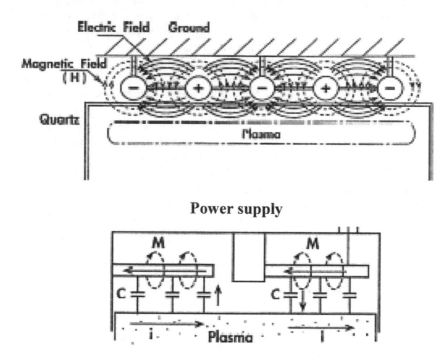

Power supply

Fig. 3. The coupling of the spoke antenna with microwave plasma [x]

(a)	(b)

Fig. 4. Images of Ar plasma at a) 80 mTorr and b) 20 Torr. The plasma maintains uniform state under a wide pressure regime.

longer than 1 m more easily than inductively coupled plasmas (ICPs). Thus, the application of MWPs to giant electronic devices such as solar cells is promising. Third, MWPs have lower bulk-electron temperature. Fourth, MWPs can be operated stably from atomic pressure down to below 10 mTorr. Fig. 4. demonstrates that Ar plasma maintains a uniform state over 22 cm in diameter up to 20 Torr. The schematic diagram of the low-pressure high-density microwave plasma utilizing the spoke antenna is shown in Fig. 5.

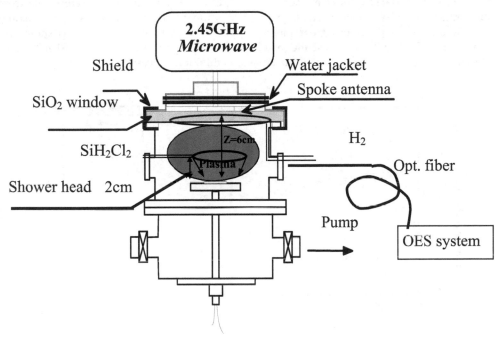

Fig. 5. The schematic diagram of the low-pressure high-density microwave plasma utilizing
the spoke antenna

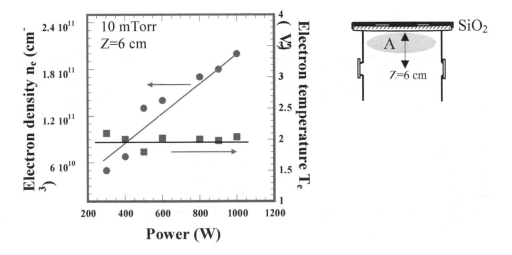

Fig. 6. Electron density, n_e, and electron temperature, T_e, measured as a function of input
microwave power.

A uniform, high-density (electron density, n_e: >10^{11} cm^{-3}) and low-temperature (electron temperature, T_e:1~2 eV) plasma can be generated by the microwave plasma source utilizing a spoke antenna without using complex components such as magnetic coil as shown in Figures 6 & 7 . The T_e is almost independent of working pressure up to ~150 mTorr as shown Figure 8, which is suitable for the large area thin film processing.

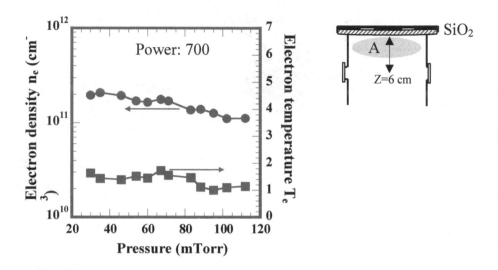

Fig. 7. Electron density and electron temperature plotted against working pressure.

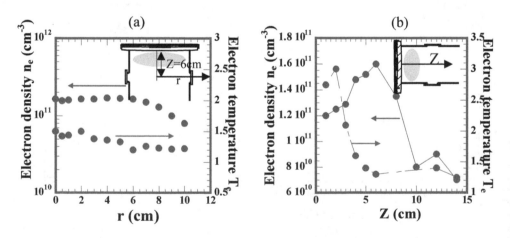

Fig. 8. The radial and axial distributions of n_e and T_e in microwave Ar plasma under microwave power of 700 W at 80 mTorr with Ar flow rate of 20 sccm.

3.2 Fast deposition of highly crystallized μc-Si:H films with low defect density from SiH₄ using low-pressure high-density microwave plasma

In this study, a new source gas supply method was introduced, i.e., the SiH$_4$ was introduced using a shower head placed 2 cm above the substrate holder under a steady flow of the H$_2$ plasma supplied by the ring. The results from these gas supply method were compared with the results from the another gas supply method, i.e. a SiH$_4$-H$_2$ mixture was fed into the chamber using a ring just beneath the quartz plate. Figure 9 shows the schematic of the two different gas supply methods. The film deposition parameters were included the SiH$_4$ concentration R=Fr(SiH$_4$)/[Fr(SiH$_4$)+Fr(H$_2$)] (Fr is the flow rate). The SiH$_4$ concentration was varied in a range from 5% to 67% by increasing Fr(SiH$_4$) from 3 to 30 sccm with a constant H$_2$ flow rate of 15 sccm. The film depositions were performed at the distance (Z) between the quartz plate and the substrate holder of 6 cm and the working pressure of 80 mTorr. The microwave power was fixed at 700W.

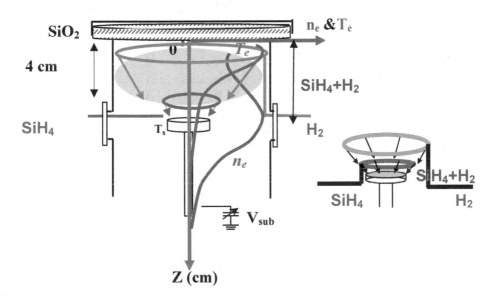

Fig. 9. Schematic of the two different gas supply methods used in this study. The distance (Z) between the quartz plate and substrate holder was 6 cm.

Fig. 10 shows the deposition rate dependence of ESR spin density, N$_s$ for the corresponding μc-Si films fabricated using two different SiH$_4$ gas supply methods at T$_s$ of 150 and 250°C. Here, the film deposition rate was controlled by varying Fr(SiH$_4$) from 3 to 30 sccm under constant Fr(H$_2$) of 15 sccm and working pressure of 80 mTorr. For all samples, the film thickness was ~ 1.5 μm and the ESR measurements were performed directly on these films. It is to be noted that N$_s$ was decreased by about one order of magnitude when the shower head was used for both T$_s$ conditions despite the other deposition conditions being the same. However, N$_s$ was almost independent of Fr(SiH$_4$) on the order of (3-4)×10^{16} cm^{-3}, which was still one order of magnitude larger than that of high quality μc-Si films reported elsewhere.

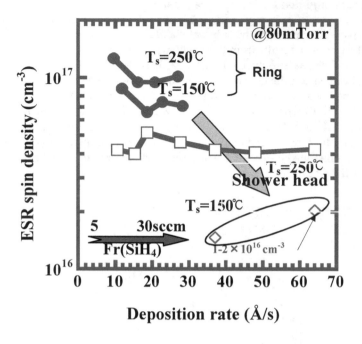

Fig. 10. ESR spin density, N_s for corresponding μc-Si films fabricated using different gas supply method as well as that for samples prepared at T_s=150°C are plotted as a function of film deposition rate R_d.

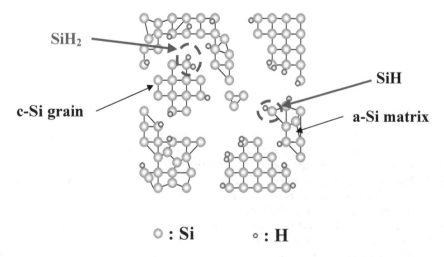

Fig. 11. μc-Si:H film microstructure

Novel Deposition Technique for Fast Growth of Hydrogenated Microcrystalline Silicon Thin-Film
for Thin-Film Silicon Solar Cells

137

A very fast deposition rate of 65Å/s has been realized for μc-Si:H films with a Raman crystallinity ratio of Ic/Ia of about 3.5 under very low H₂ dilution (i.e. with high SiH4 concentration of 67%) as shown in Fig. 12 and low defect density of (1-2) ×10¹⁶ cm⁻³ using high-density and low-temperature microwave plasma. The imaginary part of the dielectric function <ε₂> spectra of μc-Si:H films fabricated from SiH₄ using high-density and low-temperature microwave plasma is shown in Fig. 13 along with that using rf PE-CVD methods. Using the optical model the best fitted volume fraction of c-Si and void i.e. ƒc-Si and ƒvoid in the bulk layer and void in surface layer, ƒvoid, with SiH₄ concentration R for the corresponding μc-Si films is shown in Fig. 14.

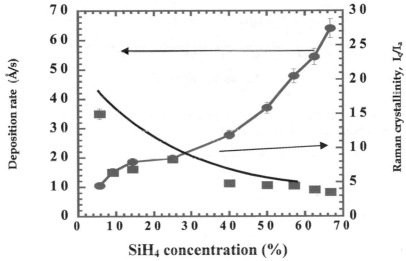

Fig. 12. Film deposition rate and Raman crystallinity, I_c/I_a as a function of SiH₄ concentration R: Fr(SiH₄)/ Fr(SiH₄)+Fr(H₂).

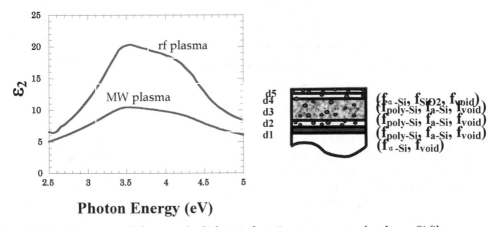

Photon Energy (eV)

Fig. 13. Imaginary part of the pseudo dielectric function<ε₂> spectra for the μc-Si films fabricated from SiH₄ using MW Plasma along with that using rf PECVD methods and five layers optical model.

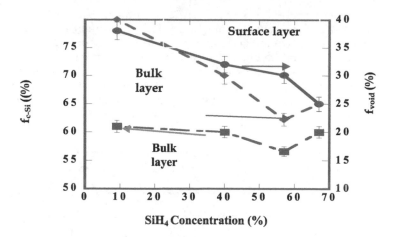

Fig. 14. Changes in f_{c-Si} and f_{void} in the bulk layer and surface layer, with SiH$_4$ concentration R for the corresponding μc-Si films shown in Fig. 11.

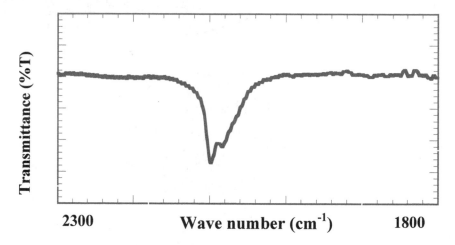

Fig. 15. The FTIR spectra for the corresponding μc-Si:H films fabricated from SiH$_4$ using MW plasma.

Highly crystallized μc-Si:H film was synthesized despite low H$_2$ dilution ratio rather than the conventional rf and VHF plasmas, because of high generation efficiency of atomic hydrogen. FTIR spectra and microstructure of μc-Si:H film and of SiH$_n$ absorption region are shown in Fig. 11 & 15 for the corresponding μc-Si film. Generally, two IR absorption peaks are observed at 2000 and 2100 cm^{-1}, which are attributed to the bulk SiH in a-Si and SiH$_2$ in μc-Si phase, respectively, in the film fabricated by the rf plasma CVD. However, no SiH absorption peak at 2000 cm^{-1} is observed in the film fabricated by high-density microwave plasma. These imply that the film crystallization is promoted extremely in the high-density plasma with negligibly small fraction of amorphous Si phase. In addition, the

Novel Deposition Technique for Fast Growth of Hydrogenated Microcrystalline Silicon Thin-Film
for Thin-Film Silicon Solar Cells

139

IR absorption peak at 2090 cm⁻¹ corresponding to the surface SiH mode in the μc-Si phase appeared as a shoulder in the high-density film. These results suggest that the c-Si phase is isolated in a-Si network as shown in Fig. 11 & 16, which is not preferable for the Si thin-film solar cells. Therefore, the suppression of the excess film crystallization is required by the selection of deposition precursor.

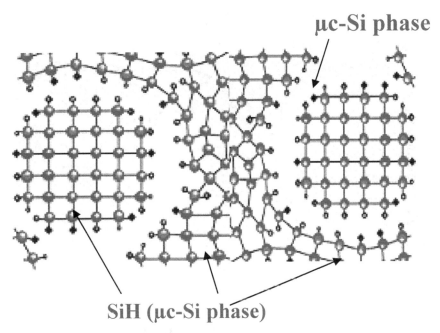

Fig. 16. μc-Si:H films fabricated from SiH₄ using MW plasma

Highly crystallized μc-Si:H films with a preferred (220) crystal orientation at a high deposition rate of 65 Å/s were fabricated from SiH₄ with a negligibly small volume fraction of amorphous Si but μc-Si network included high volume fraction of voids as shown in Fig. 15. which was hardly compatible with a device quality material. To overcome this problems, the fast deposition of highly photoconductive hydrogenated chlorinated microcrystalline Si (μc-Si:H:Cl) films with amorphous Si phase and with less volume fraction of void have been fabricated from SiH₂Cl₂ with higher threshold energy for the dissociation instead of SiH₄.

3.3 Fast deposition of highly crystallized μc-Si:H:Cl films with low defect density from SiH₂Cl₂ using low-pressure high-density microwave plasma
3.3.1 Fine structure of Si network of microcrystalline silicon thin-film fabricated from SiH₂Cl₂ and SiH₄

The typical FTIR spectra of 1-μm-thick μc-Si:H:Cl films fabricated from a SiH₂Cl₂-H₂ mixture, compared with those of μc-Si:H films from SiH₄ as shown in Fig. 17. Here, the peak assignments of SiH (bulk and surface stretching) and SiH₂ bulk stretching are also shown in Table 1.

Si-H (bulk stretching)	2000 cm⁻¹
Si-H (surface stretching)	2080 cm⁻¹
SiH₂ (bulk stretching)	2100 cm⁻¹

Table 1. Assignment of SiH, SiH₂ vibration modes

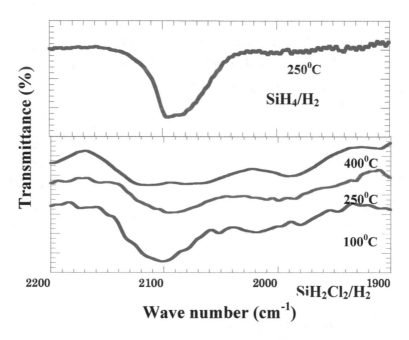

Fig. 17. FTIR spectra of µc-Si:H:Cl films at different T_ss. The typical FTIR Spectrum for the SiH$_x$ stretching absorption region in the µc- Si:H film from SiH₄ is also shown as a reference.

However, marked differences were observed in the fine structure between µc-Si:H:Cl and µc-Si:H. In the µc-Si:H films from SiH₄, the absorption peaks at 2080 cm⁻¹ and 2100 cm⁻¹ attributable to surface and bulk SiH$_x$ stretching absorption modes, respectively, in the nanocrystalline Si phase were dominant with a negligibly small peak of SiH absorption in the bulk a-Si phase at 2000 cm⁻¹. Thus, the film is highly crystallized with a negligibly small fraction of the amorphous Si phase. In addition, the IR absorption peak at 2080cm⁻¹ corresponding to the surface SiH mode in the µc-Si phase appeared as a shoulder in the µc-Si:H film. These results suggest that the c-Si phase is isolated in a-Si network, which is not preferable for the Si thin-film solar cells. Moreover, because of excess dissociation of SiH₄, the µc-Si:H network showed a porous structure, which resulted in a poor carrier transport property of photo-generated carriers.

On the other hand, both SiH and SiH₂ absorption peaks were observed at 2000 and 2100 cm⁻¹, respectively, in the µc-Si:H:Cl films, were similar as to the µc-Si:H films fabricated using conventional rf and VHF PE-CVDs of SiH₄. No SiH at the surface of µc-Si phase was observed. Moreover, the inclusion Cl in the microcrystalline Si network produces a new absorption band, which is assigned to Si-Cl bonds centered at 530cm^{-1} as described in ref. These suggest that film structure is a continuous Si network including a mixture of amorphous and crystalline Si phase, although the crystalline size is smaller. The film deposition rate reached 20 Å/s for the film synthesized from 5sccm SiH_2Cl_2, which was almost same as that for the film synthesized from SiH_4. Therefore, the fine structure of the μc-Si network from SiH_4 and SiH_2Cl_2 is different from each other.

The spectroscopy ellipsometry (SE) characterization was performed for the μc-Si:H:Cl films fabricated from a SiH_2Cl_2-H_2 mixture at different T_ss. Figure 17 shows the imaginary part of pseudo-dielectric function $<\varepsilon_2>$ spectra of μc-Si:H:Cl films fabricated from a SiH_2Cl_2-H_2 mixture at different T_ss with that of μc-Si:H from SiH_4 with a thickness of 1 μm.

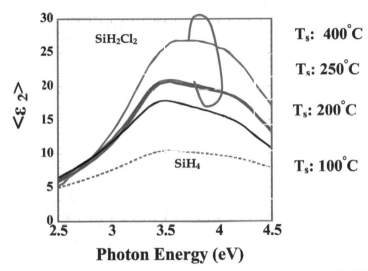

Fig. 18. Imaginary part of pseudodielectric function $<\varepsilon_2>$ spectra of μc-Si:H:Cl films fabricated from a SiH_2Cl_2-H_2 mixture at different T_ss with that of μc-Si:H from SiH_4 with a thickness of 1 μm.

In both μc-Si:H:Cl and μc-Si:H films, the fine structures were observed clearly at 3.4 and 4.2 eV, which are attributed to the E_1 and E_2 optical band transitions, respectively, in the crystalline Si (c-Si) band structure. However, the magnitude of $<\varepsilon_2>$ was much smaller in the μc-Si:H films from SiH_4 than in the μc-Si:H:Cl films from SiH_2Cl_2. Here, the magnitude of $<\varepsilon_2>$ presents qualitatively the degrees of homogeneity and the surface roughness of μc-Si films.

The $<\varepsilon_2>$ spectra were analyzed to understand the micro-structural properties of μc-Si:H:Cl films as described in section 3.2 above. The reflective index n at 2.2 eV determined by SE analysis also increased with T_s as determined using the reference poly-Si given by Jellison as shown in Fig. 19. It was higher for the μc-Si:H:Cl films in all T_s regions than that for μc-Si:H films. Thus, the rigidity of the Si-network is greater in the μc-Si:H:Cl films from SiH_2Cl_2 than in μc-Si:H films from SiH_4 using the high-density microwave plasma source.

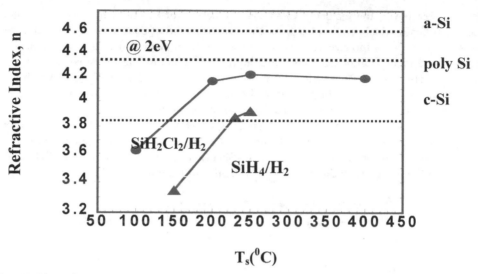

Fig. 19. The refractive index at 2.2eV in the bulk layer for μc-Si:H:Cl films plotted as a function of T_s

$f_{c\text{-}Si}$, $f_{a\text{-}Si}$ and f_{void} in the μc-Si:H:Cl films, corresponding to the bulk component, are shown in Fig.20 as a function of T_s together with those in the films synthesized from SiH_4. Notably, f_{void} in the μc-Si:H:Cl films is less than 5% despite that being 10-15% in μc-Si:H.

The differences in the fine structure of the μc-Si network between μc-Si:H:Cl films and μc-Si:H films is shown in Fig.21. The degree of the excess dissociation of SiH_2Cl_2 is considered to be suppressed rather than that of SiH_4, because the threshold energy of SiH_2Cl_2 is higher than that of SiH_4, although the high energy part of electron energy distribution (EED) also depends on the feed gas. Film crystallization was promoted efficiently in the high-density and low-temperature microwave plasma of SiH_4. However, the resulting Si film structure was still porous with much f_{void}, although $f_{c\text{-}Si}$ was over 80%. These findings originated from the excessive dissociation of both SiH_4 and H_2 in the plasma, which promoted the generation rate of not only of SiH_3 but also short life-time radicals, i.e., SiH and Si. On the other hand, $f_{a\text{-}Si}$ was still more in the μc-Si:H:Cl films than in the μc-Si:H films with less f_{void}.

3.4 Defect density of microcrystalline silicon thin-film fabricated from SiH₂Cl₂ and SiH₄

In the case of MW SiH_4 plasma, film deposition rate 65 Å/s was achieved while maintaining the low defect density but that μc-Si:H film was not available for solar cell application because of film structure was porous as described above. Similar study was performed using SiH_2Cl_2 to realize the fast deposition of μc-Si:H:Cl films with no creating additional defects, higher flux of SiH_xCl_y generated by the primary reaction in the gas phase was supplied to the depleted growing surface by increasing flow rate of SiH_2Cl_2 at a constant pressure of 120 mTorr and T_s of 250°C. Here, the deposition precursor SiH_xCl_y generated by the primary reaction in the plasma is expected to be supplied at the growing surface efficiently by increasing a flux of SiH_2Cl_2 at a constant pressure. Thus, the fast deposition of highly crystallized μc-Si:H:Cl film with lower defect density is expected because the efficient

Novel Deposition Technique for Fast Growth of Hydrogenated Microcrystalline Silicon Thin-Film
for Thin-Film Silicon Solar Cells

143

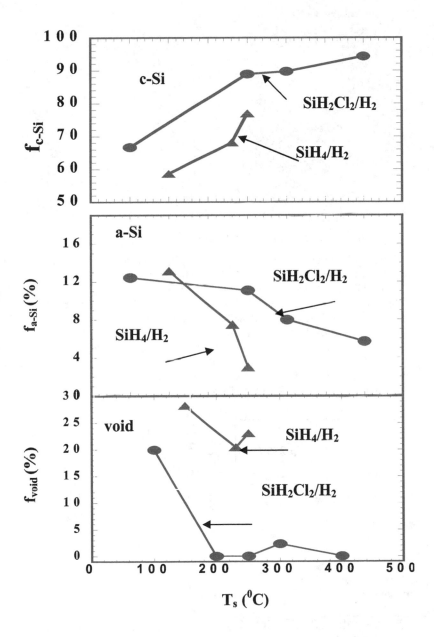

Fig. 20. The f_{c-Si}, f_{a-Si}, f_{voids} in the bulk (layer 3) of µc-Si:H:Cl films plotted as a function of T_ss. The results of µc-Si:H are also shown as a triangles symbol

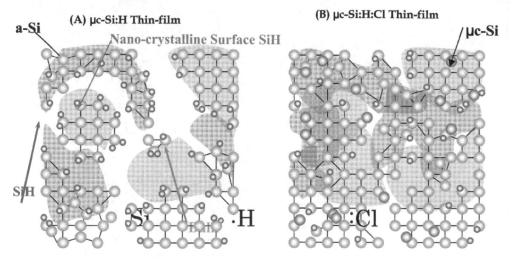

Fig. 21. Schematic of μc-Si:H and μc-Si:H:Cl network

termination of dangling bond by efficient supply of SiH_xCl_y is accelerated with the abstraction of H and Cl as HCl at the depleted growing surface. The deposition study was performed at two different T_ss of 250 and 400°C. Figure 22 demonstrates N_ss and deposition rates of μc-Si:H:Cl films fabricated at different $Fr(SiH_2Cl_2)$ at T_ss of 250 and 400°C, respectively. The high deposition rate of 40 Å/s has been achieved with increasing $Fr(SiH_2Cl_2)$ up to 20 sccm at T_ss of 400°C and 250⁰C respectively. The N_s was almost independent of $Fr(SiH_2Cl_2)$ at T_s of 250°C, whereas the N_s was markedly decreased at T_s of 400°C. These are considered because of the efficient abstraction of H and Cl at the growing surface. The film crystallization was enhanced up to flow rate of 20 sccm of SiH_2Cl_2 at T_s of 400°C. The N_ss were decreased systematically with increasing SiH_2Cl_2 at both T_ss to 4×10^{15} cm⁻³ at 15 sccm of SiH_2Cl_2.

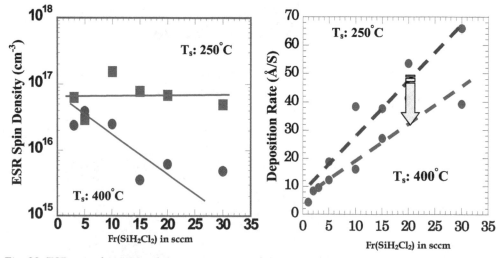

Fig. 22. ESR spin density and deposition rates of the μc-Si:H:Cl films fabricated at different flow rates of SiH_2Cl_2 at T_s of 250 and 400°C.

Novel Deposition Technique for Fast Growth of Hydrogenated Microcrystalline Silicon Thin-Film
for Thin-Film Silicon Solar Cells

145

By supplying the sufficient flux of SiH_xCl_y at a high T_s of 400°C, the termination of dangling bond is accelerated with enhancing the abstraction of H and Cl. These findings are effective to form a rigid Si network with less void fractions. In fact, the defect density N_s was almost constant of $4\text{-}5\times10^{16}$ cm^{-3} at T_s up to 250°C, whereas it decreased markedly to $3\text{-}4\times10^{15}$ cm^{-3} with $Fr(SiH_2Cl_2)$ at T_s of 400°C. Therefore, highly crystallized μc-Si:H:Cl film with low defect density was formed from a $SiH_2Cl_2\text{-}H_2$ mixture.

3.5 XRD and Raman spectra of microcrystalline silicon thin-film fabricated from SiH_2Cl_2 and SiH_4

The XRD diffraction patterns and the Raman spectrum of the μc-Si:H:Cl film fabricated at T_s of 250°C and 400°C with increasing SiH_2Cl_2 flow rates are shown in Fig 23 and Fig. 24. The XRD and Raman study of μc-Si:H:Cl films fabricated at T_s of 400°C revealed that high film crystallinities with diffraction intensities ratio of $I_{(220)}/I_{(111)}$ of 1.5-8.75 and with Raman crystallinity of I_c/I_a:5-6 were obtained.

As a consequence, highly crystallized μc-Si:H:Cl film with low defect densities of $3\text{-}4\times10^{15}$ cm^{-3} was fabricated at fast deposition rate of 27Å/s. These findings suggest that the efficient abstraction of H- and Cl- terminated growing surface.

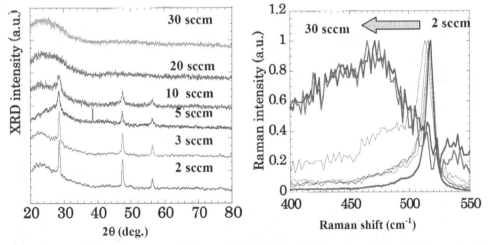

Fig. 23. XRD and Raman spectra of the μc-Si:H:Cl films fabricated at different flow rate of SiH_2Cl_2 at T_s of 250°C.

3.6 Photoelectrical properties of a-Si:H:Cl and μc-Si:H:Cl films

Fig. 25 shows the relation between the dark and photo conductivities for the μc-Si:H:Cl films fabricated by increasing the SiH_2Cl_2 flow rate at the substrate temperatures of 250°C and 400°C. The photosensitivity reached at 5-6 orders of magnitude at room temperature. The level of photoconductivity was 10^{-5} S/cm under 100 mW/cm^2 white light exposure. The dark and photo-conductivities were the order of 10^{-12} and 10^{-5} S/cm, respectively, which shows highly photosensitive films. Fig.26 shows the activation energies for the μc-Si:H:Cl films fabricated by increasing the SiH_2Cl_2 flow rate at the substrate temperatures of 250°C and 400°C The activation energies of electrical conductivity were 0.40-0.80 eV, suggesting that both a-Si:H:Cl and μc-Si:H:Cl films were intrinsic semiconductor films.

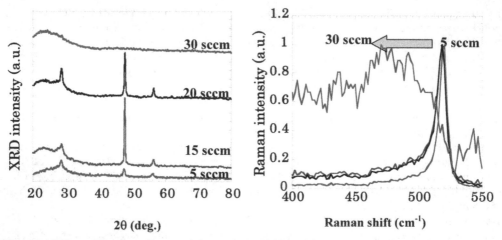

Fig. 24. XRD and Raman spectra of the μc-Si:H:Cl films fabricated at different flow rate of SiH₂Cl₂ at T$_s$ of 250 and 400°C.

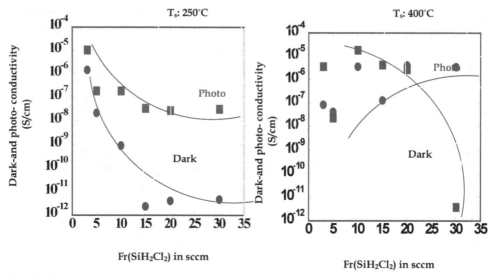

Fig. 25. Dark and photo conductivities for the μc-Si:H:Cl films as a function of SiH₂Cl₂ flow rate at the substrate temperatures of 250°C and 400°C

4. Preliminary results of p-i-n structure μc-Si:H:Cl thin-film solar cells

The preliminary result of Si thin-film solar cells using μc-Si:H:Cl thin-film fabricated by the high-density microwave plasma (MWP) of a SiH₂Cl₂-H₂ mixture are shown here. High-rate grown μc-Si:H:Cl thin-films were applied to p-i-n structure Si thin-film solar cells as intrinsic absorption layer. The solar cell was fabricated using a single chamber system. The structure of the solar cell TCO/ZnO:Al/p-i-n/ZnO:Al/Ag is as shown in Fig. 27. After the

Novel Deposition Technique for Fast Growth of Hydrogenated Microcrystalline Silicon Thin-Film for Thin-Film Silicon Solar Cells

147

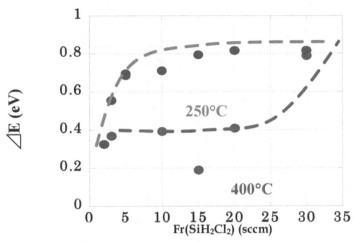

Fig. 26. Activation Energies, ΔE for the μc-Si:H:Cl films as a function of SiH_2Cl_2 flow rate at the substrate temperatures of 250°C and 400°C.

deposition of ZnO:Al and p-type Si layers on SnO_2 coated glass, respectively by rf magnetron sputtering and plasma CVD methods, μc-Si:H:Cl film with a 2-μm-thickness is fabricated using a high-density microwave plasma as a photovoltaic layer and n-type Si layer was fabricated using conventional rf plasma CVD method. When the samples were being transported between the rf chamber and MWP chamber, they were exposed to air. Subsequently, ZnO:Al and Ag layers were deposited as a top electrode using a shadow mask with a 5×5 mm² holes. Table 3 shows the typical deposition conditions for p, i and n layers, respectively. Table 4 shows the typical deposition conditions for ZnO:Al, Ag layers fabricated by rf magnetron sputtering. The photocurrent-voltage, I-V characteristics under AM 1.5, 100mW/cm² exposure condition are measured and the performance of solar cell is characterized with open circuit voltage, V_{oc}, short circuit current, I_{sc}, fill factor, FF and conversion efficiency, η. The collection efficiency was also measured from 300 to 1200 nm under bias light conditions.

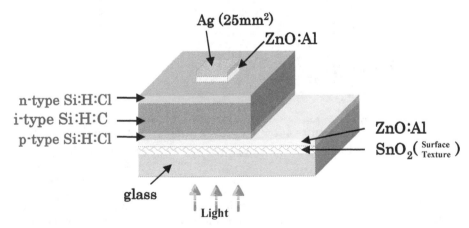

Fig. 27. The structure of the p-i-n solar cell

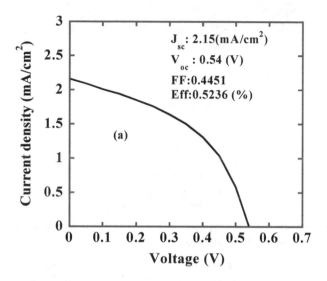

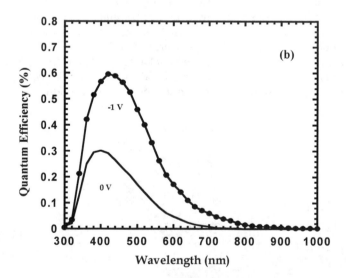

Fig. 28. (a) Photocurrent-voltage characteristics under AM 1.5 exposure condition and (b) QE spectra under -1V biased conditions of Si thin-film solar cells using a 1μm-thick μc-Si:H:Cl layer by high-density microwave plasma source.

Novel Deposition Technique for Fast Growth of Hydrogenated Microcrystalline Silicon Thin-Film
for Thin-Film Silicon Solar Cells

149

	p(RF)	i(MW)	n(RF)
Substrate Distance (mm)	20	60	20
Substrate Temperature ($\dot{o}\dot{o}$C)	250	250	250
Power (W)	5	700	5
Pressure (mTorr)	200	120	200
H_2 (sccm)	61	15	55
SiH_2Cl_2 (sccm)	3	3	3
PH_3 (sccm)			15
B_2H_6 (sccm)	9		

Table 3. Typical deposition conditions for p, i and n layers

	ZnO:Al(front side)	ZnO:Al(back side)	Ag
Substrate Position (cm)	6	6	6
Substrate Temperature (°C)	350	250	250
RF Power (W)	100	100	50
Pressure (Pa)	2.5	2.5	2.5
thickness (Å)	2500	2500	1500

Table 4. Typical deposition condition for ZnO:Al and Ag layers

Fig.28 illustrates photocurrent –voltage characteristics for Si:H:Cl thin-film solar cells under 100 mW/cm² white light exposure. Fig. 28a shows the I-V characteristics for the cell using μc-Si:H:Cl films fabricated at 20 Å/s by the high-density microwave plasma CVD of SiH_2Cl_2. The 5-6% efficiencies have been achieved for the cells fabricated by the conventional rf plasma-CVD method. However, the performance is still poor and the open circuit voltage, (Voc):0.54 V, short circuit density, (Jsc):2.15 mA/cm², Fill Factor, FF:0.5236 and the conversion efficiency was 0.5236% in the cell made by the high-density microwave plasma from SiH_2Cl_2 but solar cell performance is confirmed by the high-density microwave plasma from SiH_2Cl_2 for the first time. The diffusion of Boron and Chlorine happens easily in i-layer by the high-density microwave plasma. Moreover, the etching reaction of p layer has occurred because of the hydrogen plasma. It is required to evaluate not only a single film but it is also necessary to evaluate the each interface i.e. AZO/p, p/i and i/n in order to improve the solar cell performance. More over precise control of p/i, i/n, AZO/p interface formation is needed for obtaining the further high performance.

5. Conclusion

The highly photoconductive and crystallized μc-Si:H:Cl films with less volume fraction of void and defect density were synthesized using the high-density and low-temperature microwave plasma source of a SiH_2Cl_2-H_2 mixture rather than those from SiH_4 while maintaining a high deposition rate of 27 Å/s. The μc-Si:H:Cl film possesses a μc-Si and a-Si mixture structure with less volume fraction of voids. The role of chlorine in the growth of μc-Si:H:Cl films is the suppression of the excess film crystallization at the growing surface. H termination of growing surface is more effective to suppress the defect density rather than that of Cl termination. The fast deposition of the μc-Si:H:Cl film with low defect density of 3-4 ×10¹⁵ cm⁻³ is achieved with reducing Cl concentration during the film growth. Both a-Si:H:Cl and μc-Si:H:Cl films show high photoconductivity of 10^{-5} S/cm under 100 mW/cm^{-2} exposure, are the possible materials for Si thin-film solar cells. The performance of p-i-n solar cell from μc-Si:H:Cl films using the high-density microwave plasma source was confirmed for the first time.

6. References

Ziegler, Y. (2001),More stable low gap a-Si:H layers deposited byPE-CVD at moderately high temperature with hydrogen dilution". *Solar Energy Materials & Solar Cells, 2001.* 66: p. 413- 419.

Graf S. (2005). *"Single-chamber process development of microcrystalline Silicon solar cells and high-rate deposited intrinsic layers"*, in institute de Microtechnique, Universite de Neuchatel: Neuchatel.

Meillaud, F.(2005). "Light-induced degradation of thin-film microcrystalline silicon solar cells". in *31th IEEE Photovoltaic Specialist Conference.* 2005, Lake Buena Vista, FL, USA

Veprek, S. and V. Marecek (1968). "The preparation of thin layers of Ge and Si by chemical hydrogen plasma transport", *Solid State Electronics, 1968,* Vol. 11: p. 683-684.

LeComber, P.G. and W.E. Spear (1970). *"PECVD: plasma enhanced chemical vapor deposition".* *Physical Review Letters,* 1970. Vol. 25: p. 509.

A. Madan, S. R. Ovshinsky and E. Benn (1979). *Phil Mag.* B 40 (1979) 259

B. Chapman: Glow Discharge Processes. *Sputtering and Plasma Etching,* Chapter 9, John Wiley

A. Bogaerts, E. Neyts, R. Gijbels, J. van der Mullen(2002). *Spectrochimica Acta* B 57 (2002) 609J.K. Saha, N. Ohse, H. Kazu, Tomohiro Kobayshi and Hajime Shirai (2007). *"18th International Symposium on Plasma Chemistry proceedings"*, Kyoto, Japan, Aug 26-31, 2007.

J. K. Saha, Naoyuki Ohse, Hamada Kazu, Tomohiro Kobayshi and Hajime Shirai (2007). *Japan Society of Applied Physics and Related Societies (the 54th Spring Meeting),* Aoyama Gakuin University, March 27-March, 30,2007, 27p-M-2

S. Samukawa, V. M. Donnelly and M. V. Malyshev (2000). *Jpn. J. Appl. Phys.* 39 (2000) 1583

I. Ganachev and H. Sugai (2007). *Surface and Coating Technology* 174-175 (2003) 15.

J. K. Saha, H. Jia, N. Ohse and H. Shirai(2007). *Thin Solid Films* 515 (2007) 4098.

Y.Nasuno, M.Kondo and A. Matsuda (001). *Tech. Digest of PVSEC-12.* Jeju, Korea, 2001,791.

L. Guo, Y. Toyoshima, M.Kondo and A. Matsuda (1999). *Appl. Phys. Lett.* 75 (1999) 3515

G. E. Jellison, Jr. (1992). Opt. Mater. 1 (1992) 41

S. Kalem, R. Mostefaoui, J. Chevallier (1986). *Philos. Mag.* B 53 (1986) 509-513.

J.K. Saha, N. Ohse, K. Hamada, T. Kobayshi, H.Jia and H. Shirai (2010). *Solar Energy Materials & Solar Cells* 94 (2010) 524- 530.

J. K. Saha, N. Ohse, K. Hamada, K. Haruta, T.Kobayshi, T. Ishikawa, Y. Takemura and H. Shirai (2007). *Jpn. J. Appl. Phys.* 46 (2007) L696.

D.E. Aspnes (1976). Spectroscopic ellipsometry of solids, in: B.O. Seraphin (Ed.), *Optical Properties of Solids: New Developments*, North-Holland, Amsterdam, 1976, pp. 801–846 (Chapter 15).

Hiroyuki Fujiwara (2007). *Spectroscopic Ellipsometry: Principles and Applications*, John Wiley & Sons, Ltd., 2007, pp. 189–191.

H. Kokura and H. Sugai (2000). *Jpn. J. Appl. Phys.* 39 (2000) 2847

J. K. Saha, N. Ohse, K. Hamada, T.Kobayshi and H. Shirai(2007). *Tech. Digest of PVSEC-17.* Fukuoka, Japan, 2007, 6P-P5-68.

Y. Li, Y. Ikeda, T. Saito and H. Shirai (2006). *Thin Solid Films,* 511-512(2006) 46

8

Development of Flexible Cu(In,Ga)Se$_2$ Thin Film Solar Cell by Lift-Off Process

Yasuhiro Abe, Takashi Minemoto and Hideyuki Takakura
Ritsumeikan University
Japan

1. Introduction

Clean energy resources as an alternative to fossil fuels has been required. Photovoltaics is the most promising among renewable energy technologies. On the other hand, the cost of the electrical energy generated by the solar cells was higher than that generated by fossil fuels. The cost reduction of the solar cell is therefore required.

Since high-conversion efficiencies have been demonstrated for solar cells using GaAs substrates in 1977 (Kamath et al., 1977; Woodall et al., 1977), a critical problem is how to reduce power generation cost. The characters required to solar cells strongly depend on its applications. In particular, thin film solar cells are promising for terrestrial applications, because thin film solar cells are more advantageous than bulk type solar cells in terms of consumption of raw materials. Konagai et al. fabricated the thin film solar cells on a single crystalline GaAs substrate by the liquid phase epitaxy mehtod, and focused on the reuse of GaAs substrates by detaching these thin film solar cells from the GaAs substrates (Konagai et al., 1978). Konagai et al. named this separation technique the Peeled Film Technology (PFT). This is the invention of the lift-off method in solar cell development. A specific explanation of the PFT is as follows. An Al$_{1-x}$Ga$_x$As layer was introduced between the thin film solar cell and the GaAs substrate as a release layer. The thin film solar cell was separated from the GaAs substrate by etching the Al$_{1-x}$Ga$_x$As layer by the HF solution, because Al$_{1-x}$Ga$_x$As was readily dissolved by the HF solution compared to GaAs. Since a chemical technique was mainly used for the peeling, this method is defined as a chemical lift-off process. Recently, this has been researched as the epitaxial lift-off (ELO) method (Geelen et al., 1997; Schemer et al., 2000, 2005a; Voncken et al., 2002; Yablonovitch et al., 1987).

On the other hand, the cleavage of lateral epitaxial films for transfer (CLEFT) process, where the thin film was mechanically peeled, was developed as a transfer method of a single crystalline GaAs thin film (McClelland et al., 1980). A specific explanation of the CLEFT process is as follows. A photoresist was applied to a surface of a GaAs substrate. The photoresist was patterned with equally-spaced stripe openings by standard photolithographic techniques. Next, a GaAs layer was grown on this patterned substrate surface. In this case, a GaAs layer was grown on only the openings of the photoresist. The lateral growth of a GaAs layer occurs during the GaAs deposition. A single crystalline GaAs layer is therefore formed on the photoresist. Alternative substrate was bonded onto this surface with epoxy glue. The single crystalline thin film was transferred to the alternative substrate by applying tensile strain. The CLEFT process is theorefore defined as a mechanical lift-off process.

Unfortunately, the conversion efficiency of the GaAs thin film solar cell using the lift-off process was lower than that of the GaAs bulk solar cell (Schermer et al., 2006). Recently, comparable conversion efficiencies have been demonstrated (Bauhuis et al., 2009).

On the other hand, the energy:weight ratio of the photovoltaic module is a very important index for space applications. Integration of high-efficiency III-V solar cells with light weight substrates is required. Schermer et al. developed high-efficiency III-V solar cells with light-weight by the ELO process using the GaAs substrate (Schermer et al., 2005b).

In addition, the lift-off process was applied to reuse Si substrates (Bergmann et al., 2002; Brendel, 2001). Moreover, the lift-off process was applied to fabricate flexible solar cells in the developments of II-VI and I-III-VI$_2$ semiconductor thin film solar cells (Marrón et al., 2005; Minemoto et al., 2010; Romeo et al., 2006; Tiwari et al., 1999).

Here, we focus on advantages of the lift-off process in fabrication of flexible Cu(In,Ga)Se$_2$ (CIGS) thin film solar cells. For example, for the fabrication process where CIGS layers were directly grown on flexible substrates, Ti foils (Hartmann et al., 2000; Herz et al., 2003; Ishizuka et al., 2009a; Kapur et al., 2002; Kessler et al., 2005; Yagioka & Nakada, 2009), Cu steel sheets (Herz et al., 2003), Mo foils (Kapur et al., 2002, 2003), stainless steel sheets (Britt et al., 2008; Gedhill et al., 2011; Hashimoto et al., 2003 ; Kessler et al., 2005; Khelifi et al., 2010; Pinarbasi et al., 2010; Satoh et al., 2000, 2003; Shi et al., 2009; Wuerz et al., 2009), Al foils (Brémaud et al., 2007), Fe/Ni alloy foils (Hartmann et al., 2000), ZrO$_2$ sheets (Ishizuka et al., 2008a, 2008b, 2009b, 2010), and polyimide (PI) films (Brémaud et al., 2005; Caballero et al., 2009; Hartmann et al., 2000; Ishizuka et al., 2008c; Kapur et al., 2003; Kessler et al., 2005; Rudmann et al., 2005; Zachmann et al., 2009;), are used as flexible substrates. Since these materials do not include Na, other processes to introduce Na are required (Caballero et al., 2009; Ishizuka et al., 2008a; Keyes et al., 1997). Since the thermal tolerance temperature of a PI film is ~450°C, the low temperature growth of a CIGS layer is required. The first of the advantages of the lift-off process is to enable to use a high quality CIGS layer grown on a conventional Mo/soda-lime glass (SLG) substrate in the flexible solar cell fabrication. Consequently, the low temperature growth technology for high quality CIGS layer formation and novel processes for a Na source are not required. The second is to enable to use low thermal tolerance films as the flexible substrate of a CIGS solar cell, because in the CIGS solar cell fabrication process, the highest temperature process is the growth of a CIGS layer and the process temperature after the growth of a CIGS layer is less than 100°C.

2. Experimental

2.1 Flexible Cu(In,Ga)Se$_2$ solar cell fabrication procedure

A schematic illustration of the fabrication procedure of our flexible CIGS solar cell using the lift-off process is shown in Fig. 1 (Minemoto et al., 2010). A 0.8-μm-thick Mo layer was deposited on an SLG substrate without intentional substrate heating by the radio frequency (RF) magnetron sputtering method. A 2.5-μm-thick CIGS layer was deposited on the Mo/SLG substrate by the three-stage deposition process at the highest substrate temperature of approximately 550°C (Contreras et al., 1994a; Negami et al., 2002). From energy dispersive x-ray spectrometry measurements, the Cu, In, Ga, and Se composition ratios of this CIGS layer were approximately 23, 18, 8, and 51%, respectively. The

[Cu]/[Ga+In] and [Ga]/[Ga+In] ratios of the CIGS layer were therefore calclated to be ~0.88 and ~0.31, respectively. After CIGS surface cleaning by a KCN solution, a 0.2-μm-thick Au layer was deposited on the CIGS surface by a resistive evaporation method as a back electrode. The samples were annealed for 30 min at 250°C in N$_2$ ambient. Flexible films were bonded onto support SLG substrates with a silicone adhesion bond for preparation of the alternative substrates. These alternative substrates were also bonded onto the Au/CIGS/Mo/SLG structure with conductive epoxy glue. To dry the conductive epoxy glue, the samples were annealed on a hot plate at 100°C for 10 min in the atmosphere. Then, the alternative-sub./epoxy/Au/CIGS stacked structures were detached from the primary Mo/SLG substrates by applying tensile strain. In this detachment, the CIGS layer was transferred to the alternative substrate side (Marrón et al., 2005). The lift-off flexible CIGS solar cells were fabricated using this peeled CIGS layer. After cleaning of the CIGS rear surface by a KCN solution, a 0.1-μm-thick CdS layer was deposited on the CIGS rear surface by the chemical bath deposition method. 0.1-μm-thick i-ZnO and 0.1-μm-thick In$_2$O$_3$:Sn layers were deposited by the RF magnetron sputtering method. Al/NiCr grids were formed. Finally, the flexible CIGS solar cells using the lift-off process were completed by detaching the flexible films from the support SLG substrates. For comparison, we also prepared a standard solar cell where the lift-off process was not carried out (the Al/NiCr/In$_2$O$_3$:Sn/ZnO/CdS/CIGS/Mo/SLG structure). The properties of the films used in this study are summarized in Table 1. Figure 2 shows a photograph of the flexible solar cells using the PI film.

Material	PI	PTFE	Polyester
Thermal tolerance temperature (°C)	450	260	120
Thickness (μm)	55	120	25

Table 1. Properties of PI, polytetrafluoroethylene (PTFE) and polyester films used in this study are summarized.

2.2 Characterization methods

Current density-voltage (J-V) measurements were performed under standard air mass 1.5 global conditions (100 mW/cm^2) at 25°C. External quantum efficiency (EQE) measurements of the AC mode were performed at 25°C under white light bias (~0.3 sun) conditions. The laser-beam-induced current (LBIC) method using the laser diode (λ: 783 nm, laser power: 0.3 mW) was performed to investigate a spatial distribution of an EQE (Minemoto et al., 2005). In LBIC measurements, a nominal spot size is less than 50 μm and a scan step is 53 μm. The surfaces of the fabricated flexible solar cells were observed with an optical microscope. The J-V, EQE, and LBIC measurements were performed after light soaking.

3. Results and discussion

3.1 Characterization of flexible Cu(In,Ga)Se$_2$ solar cells fabricated using lift-off process

The J-V characteristics of the fabricated flexible solar cells are shown in Fig. 3. For comparison, the J-V characteristic of the standard solar cell is also shown. Solar cell parameters such as the short-circuit current density (J_{sc}), the open-circuit voltage (V_{oc}), the

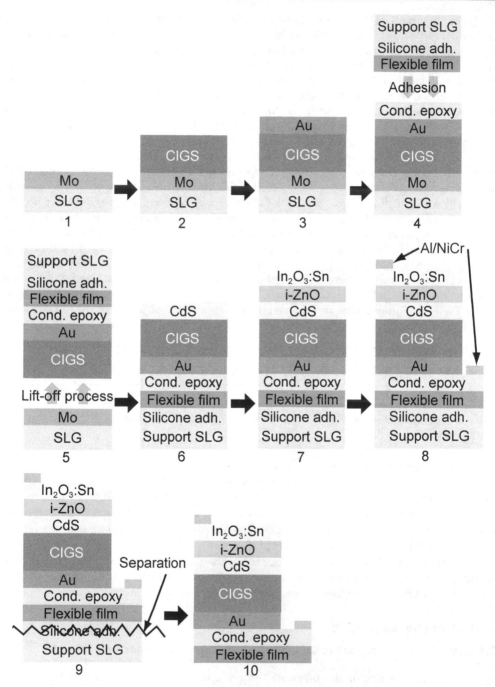

Fig. 1. Schematic illustration of fabrication procedure of flexible CIGS solar cell using lift-off process.

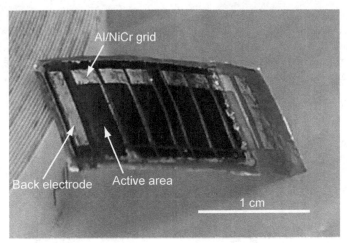

Fig. 2. Photograph of flexible CIGS solar cells using PI film.

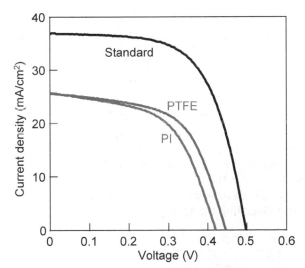

Fig. 3. Photo J-V curves of flexible solar cells using PTFE (red) and PI (blue) films. Photo J-V curve of standard solar cell without lift-off process (brack) is also shown for comparison.

Sample structure	Eff. (%)	J_{sc} (mA/cm²)	V_{oc} (V)	FF (%)
PI flexible	5.9	25.7	0.420	54.9
PTFE flexible	6.6	25.6	0.445	57.9
Standard	11.4	36.9	0.497	62.4

Table 2. Solar cell parameters obtaind from flexible solar cells using PI and PTFE films. Solar cell parameters of standard solar cell are also shown for comparison.

conversion efficiency (*Eff.*), and the fill factor (*FF*) are summarized in Table 2. The conversion efficiencies of the flexible solar cells are an approximately half conversion efficiency of the standard solar cell. EQE spectra of these solar cells are shown in Fig. 4. EQEs of the flexible solar cells remarkably decrease in the long wavelength region from 700 to 1200 nm compared to the standard solar cell. We discuss this cause as below.

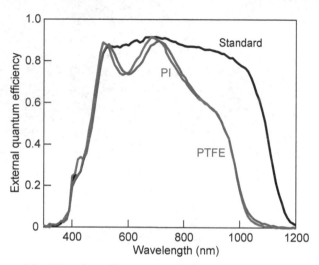

Fig. 4. EQE spectra of flexible solar cells using PTFE (red) and PI films (blue). EQE spectrum of standard solar cell without lift-off process (black) is also shown for comparison. EQE spectra of flexible solar cells are similar irrespective of substrate materials.

As shown in Fig. 5(a), the band gap profile of the standard solar cell consists of the graded band gap structure because of the three-stage deposition process. The diffusion length of electrons generated by the long wavelength light near the back electrode is improved due to the quasi-electric field in which the CIGS layer forms (Contreras et al., 1994b). The graded band gap structure is therefore beneficial for collecting the photogenerated carriers. On the other hand, as shown in Fig. 5(b), the band gap profile of the CIGS layer is inverted due to

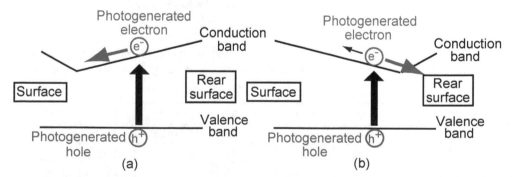

Fig. 5. Schematic illustrations of band gap profiles of CIGS layers. CIGS absorber layers with (a) double geraded band gap and (b) inverted double graded band gap structures are shown.

the lift-off process for the flexible solar cells. We speculate that the band gap profile of the inverted graded band gap structure is not beneficial for collecting the photogenerated carriers by long wavelength light. We conclude that the EQE reductions observed for the flexible solar cells are attributed to the influence of the inverted graded band gap structure.

We describe an interesting point of our flexible solar cells as below. Different materials with different thermal tolerance temperatures are used as the flexible substrates of these flexible solar cells, as shown in Table 1. These flexible solar cells, however, show the similar characteristics irrespective of the flexible film materials from Fig. 3 and Fig. 4.

LBIC and optical microscope images of the flexible solar cell using the PTFE film are shown in Figs. 6(a) and 6(b), respectively. There is a low EQE region on the lower side of the solar cell from Fig. 6(a). This low EQE region corresponds approximately to the flexurelike region from a comparison between Figs. 6(a) and 6(b). This result therefore suggests that this flexure cause reduction of an EQE. LBIC and optical microscope images of the standard solar cell are shown in Figs. 6(c) and 6(d), respectively. In contrast, the LBIC and optical microscope images are uniform for the standard solar cell.

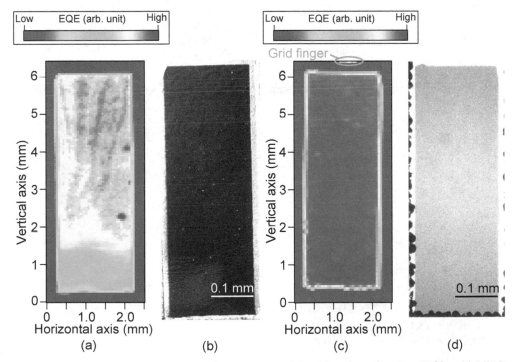

Fig. 6. (a) LBIC and (b) optical microscope images of flexible solar cell using PI film. (c) LBIC and (d) optical microscope images of standard solar cell. Indicators of EQE intensity are shown next to LBIC images.

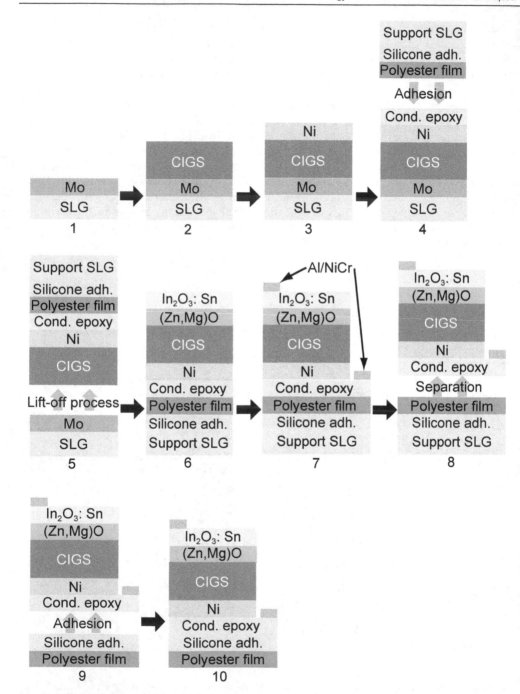

Fig. 7. Schematic illustration of fabrication procedure of flexible CIGS solar cell using $(Zn_{0.83}, Mg_{0.17})O$ window layer and lift-off process.

3.2 Development of Cd-free flexible Cu(In,Ga)Se$_2$ solar cells

We developed a new Cd-free flexible CIGS solar cell using a (Zn,Mg)O window layer. The fabrication procedure is shown in Fig. 7. This process is basically similar to Fig. 1. We deposited a 0.1-μm-thick $(Zn_{0.83},Mg_{0.17})O$ window layer in stead of the ZnO window/CdS buffer layers. The RF magnetron cosputtering method using ZnO and MgO targets was used as the deposition technique (Minemoto et al., 2000, 2001). We also deposited a 0.2-μm-thick Ni layer by the resistive evaporation method as the back electrode in stead of the Au layer. In this subsection, a 55-μm-thick polyester film was used as a flexible substrate. Interestingly, when the flexible solar cell using the polyester film was separated from the support SLG substrate, the detachment occurred not at the support SLG/polyester interface but at the polyester/epoxy interface due to the weaker adhesion at the polyester/epoxy interface. After the substrate-free structure was once, the polyester film was therefore bonded onto the rear surface of the solar cell with a silicone adhesion bond. The photograph of the flexible solar cells fabricated via the above procedure is shown in Fig. 8. We also prepared not only the flexible solar cells using the conventional ZnO window/CdS buffer layers but also the solar cells without the lift-off process for comparison.

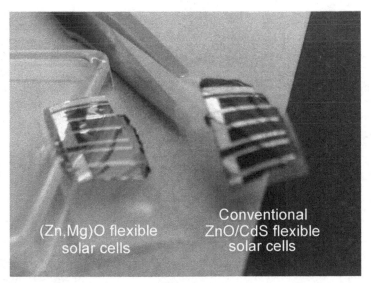

Fig. 8. Photograph of flexible solar cells using polyester film. Left solar cells are Cd-free solar cells using (Zn,Mg)O window layer. Right solar cells consist of conventional ZnO window/CdS buffer layers structure.

The *J-V* characteristics of the flexible solar cells are shown in Fig. 9. The results of the standard solar cells without the lift-off process are also shown in Fig. 9. Solar cell parameters obtained from the *J-V* characteristics are summarized in Table 3. All parameters of the ZnO/CdS solar cell is higher than those of the (Zn,Mg)O solar cell for the standard solar cells. On the other hand, although there are the differences in the window layer/ buffer layer structures for the flexible solar cells, these flexible solar cells show the similar properties.

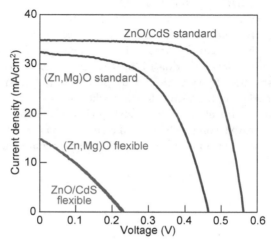

Fig. 9. Photo *J-V* curves of flexible solar cells using (Zn,Mg)O window layer and conventional ZnO window/CdS buffer layers. Photo *J-V* curves of standard solar cells without lift-off process are also shown for comparison.

EQE spectra of these solar cells are shown in Fig. 9. EQEs of the (Zn,Mg)O standard solar cell are higher than those of the ZnO/CdS standard solar cell in the region from 300 to 480 nm, because the band gap of $(Zn_{0.83},Mg_{0.17})O$ is higher than those of CdS and ZnO (Minemoto et al., 2000). These high EQEs in this region is therefore attributed to a low transmission loss of the short wavelength light. Moreover, the tendency of this result is also observed for the flexible solar cells. We found that the (Zn,Mg)O window layer structure was effective for reducing a transmission loss of the short wavelength light even in our flexible solar cells.

Sample structure	*Eff.* (%)	J_{sc} (mA/cm²)	V_{oc} (V)	*FF* (%)
(Zn,Mg)O flexible	1.0	14.8	0.231	30.5
ZnO/CdS flexible	1.0	14.8	0.227	30.2
(Zn,Mg)O standard	8.3	32.4	0.465	54.9
ZnO/CdS standard	13.7	34.9	0.562	70.0

Table 3. Summary of solar cell parameters obtained from flexible solar cells using (Zn,Mg)O window layer and conventinal ZnO window/CdS buffer layers. For comparison, solar cell parametaers otained from standard solar cells using (Zn,Mg)O window layer and ZnO window/CdS buffer layers are also summarized.

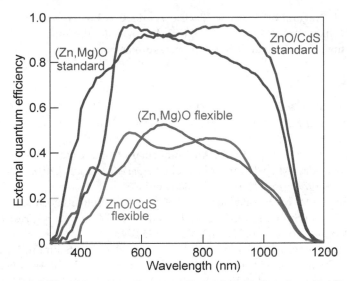

Fig. 10. EQE spectra of flexible solar cells using (Zn,Mg)O window layer (red) and conventional ZnO window/CdS buffer layers (blue). EQE spectra of standard solar cells using (Zn,Mg)O window layer (dark red) and ZnO window/CdS buffer layers (dark blue) are also shown for comparison.

Here, we discuss why these flexible solar cells showed the similar solar cell parameters. In this subsection, we used Ni in stead of Au as a back electrode material. In subsection 3.1, the ZnO/CdS flexible solar cells with the Au back electrode showed a conversion efficiency of ~6%. We think that the Ni back electrode may limit performance of these solar cells. We therefore speculate that the Ni atoms, which diffused into the CIGS layer from the back side due to the low temperature annealing, behave as recombination centers for electrons.

4. Conclusion

After we described the review of the lift-off process, we also described the advantages of the lift-off process in the flexible CIGS solar cell fabrication. We developed the fabrication procedure of the flexible CIGS solar cells using the lift-off process. The characteristics of the flexible solar cells were shown compared to the standard solar cell. Although the conversion efficiencies of the flexible solar cells using the lift-off process are an approximately half conversion efficiency of the standard solar cell, the flexible solar cells showed the similar characteristics irrespective of the substrate materials. Moreover, we attempted the concept of a Cd-free solar cell. We found that the choice of back electrode materials is a crucial problem rather than the window layer/buffer layer structure. We expect that the lift-off process further advances through our results.

5. Acknowledgment

This work was partially supported by the Ministry of Education, Culture, Sports, Science and Technology (MEXT) through a Grant-in-Aid for Young Scientists (B). The authors are grateful to Dr. T. Negami of Pnasonic Electric Works Co., Ltd., for useful discussion. The authors would like to thank Mr. T. Yagi and Associate Professor S. Ikeda of Osaka University for their technical support in EQE measurements.

6. References

Kamath, G. S.; Ewan, J. & Knechtli R. C. (1977). Large-Area High-Efficiency (AlGa)As-GaAs Solar Cells. *IEEE Transactions on Electron Devices*, Vol. ED-24, No. 4, (April 1977), pp. 473-475, ISSN 0018-9383

Woodall, J. M. & Hovel, H. J. (1977). An isothermal etchback-regrowth method for high efficiency $Ga_{1-x}Al_xAs$-GaAs solar cells. *Applied Physics Letters*, Vol. 30, No. 9, (February 1977), pp. 492-494, ISSN 0003-6951

Konagai, M.; Sugimoto, M. & Takahashi, K. (1978). HIGH EFFICIENCY GaAs THIN FILM SOLAR CELLS BY PEELED FILM TECHNOLOGY. *Journal of Crystal Growth*, Vol. 45, (December 1978), pp. 277-280, ISSN 0022-0248

Geelen, A. V.; Hageman, P. R. Bauhuis, G. J. Rijsingen, P. C. V. Schmidt, P. & Giling, L. J. (1997). Epitaxial life-off GaAs solar cell from a reusable GaAs substrate. *Material Science and Engineering B*, Vol. 45, (March 1997), pp. 162-171, ISSN 0921-5107

Schermer, J. J.; Bauhuis, G. J. Mulder, P. Meulemeesters, W. J. Haverkamp, E. Voncken, M. M. A. J. & Larsen, P. K. (2000). High rate epitaxial lift-off on InGaP films from GaAs substrate. *Applied Physics Letters*, Vol. 76, No. 15, (April 2000), pp. 2131-2133, ISSN 0003-6951

Schermer, J. J.; Mulder, P. Bauhuis, G. J. Voncken, M. M. A. J. Deelen, J. V. Haverkamp, E. & Larsen, P. K. (2005a). Epitaxial Lift-Off for large area thin film III/V devices. *Physica Status Solidi A*, Vol. 202, (Febrary 2005), pp. 501-508, ISSN 1862-6300

Voncken, M. M. A. J.; Schermer, J. J. Maduro, G. Bauhuis, G. J. Mulder, P. & Larsen. P. K. (2002). Influence of radius of curvature on the lateral etch rate of the weight induced epitaxial lift-off process. *Materials Science and Engineering B*, Vol. 95, (September 2002) pp. 242-248, ISSN 0921-5107

Yablonovitch, E.; Gmitter, T. Harbison, J. P. & Bhant, R. (1987). Extreme selectivity in the lift-off epitaxial GaAs films. *Applied Physics Letters*, (December 1987), pp. 2222-2224, ISSN 0003-6951

McClelland, R. W.; Bozler, C. O. & Fan, J. C. C. (1980). A technique for producing epitaxial films on reusable substrate. *Applied Physics Letters*, Vol. 37, No. 6, (September 1980), pp. 560-562, ISSN 0003-6951

Schermer, J. J.; Bauhuis, G. J. Mulder, P. Haverkamp, E. J. Deelen, J. V. Niftrik, A. T. J. V. & Larsen, P. K. (2006). Photon confinement in high-efficiency thin-film III-V solar cells obtained by epitaxial lift-off. *Thin Solid Films*, Vol. 511-512, (January 2006), pp. 645-653, ISSN 0040-6090

Bauhuis, G. J.; Mulder, P. Haverkamp, E. J. Huijben, J. C. C. M. & Schermer, J. J. 26.1% thin-film GaAs solar cell using epitaxial lift-off. *Solar Energy Materials and Solar Cells*, Vol. 93, (May 2009), pp. 1488-1491, ISSN 0927-0248

Schermer, J. J.; Mulder, P. Bauhuis, G. J. Larsen, P. K. Oomen G. & Bongers E. (2005b) Thin-film GaAs Epitaxial Lift-off Solar Cells for Space Applications. *Progress in Photovoltaics: Research and Applications*, Vol. 13, (April 2005), pp. 587-596, ISSN 1062-7995

Bergmann, R. B.; Berge, C. Rinke, T. J. Schmidt, J. & Werner, J. H. (2002). Advances in monocrystalline Si thin film solar cells by layer transfer. *Solar Energy Materials and Solar Cells*, Vol. 74, (October 2002), pp. 213-218, ISSN 0927-0248

Brendel, R. (2001). Review of Layer Transfer Processes for Crystalline Thin-Film Silicon Solar Cells. *Japanese Journal of Applied Physics*, Vol. 40, No. 7, (July 2001), pp. 4431-4439, ISSN 0021-4922

Marrón, D. F.; Meeder, A. Sadewasser, S. Würz, R. Kaufmann, C. A. Glatzel, T. Schedel-Niedrig, T. & Lux-Steiner, M. C. Lift-off process and rear-side characterization of CuGaSe₂ chalcopyrite thin films and solar cells. *Journal of Applied Physics*, Vol. 97, (April 2005), pp. 094915-1-094915-7, ISSN 0021-8979

Minemoto, T.; Abe, Y. Anegawa, T. Osada, S. & Takakura, H. (2010). Lift-Off Process for Flexible Cu(In,Ga)Se₂ Solar Cells. *Japanese Journal of Applied Physics*, Vol. 49, No. 4, (April 2010), pp. 04DP06-1-04DP06-3, ISSN 0021-4922

Romeo, A.; Khrypunov, G. Kurdesau, F. Arnold, M. Bätzner, D. L. Zogg, H. & Tiwari, A. N. High-efficiency flexible CdTe solar cells on polymer substrates. *Solar Energy Materials and Solar Cells*, Vol. 90, (November 2006), pp. 3407-3415, ISSN 0927-0248

Tiwari, A. N.; Krejci, M. Haung, F.-J. & Zogg, H. (1999). 12.8% Efficiency Cu(In,Ga)Se₂ Solar Cell on a Flexible Polymer Sheet. *Progress in Photovoltaics: Research and Applications*, Vol. 7, (October 1999), pp. 393-397, ISSN 1062-7995

Hartmann, M.; Schmidt, M. Jasenek, A. Schock, H. W. Kessler, F. Herz, K. & Powalla, M. Flexible and light weight substrates for Cu(In,Ga)Se₂ solar cells and modules, *Conference Record of the Twenty-Eighth IEEE Photovoltaic Specialists Conference 2000*, pp. 638-641, ISBN 0-7803-5772-8, Anchorage, Alaska, USA, September 15-22, 2000

Herz, K.; Eicke, A. Kessler, F. Wächter, R. & Powalla, M. (2003). Diffusion barriers for CIGS solar cells on metallic substrates. *Thin Solid Films*, Vol. 431-432, (May 2003), pp. 392-397, ISSN 0040-6090

Ishizuka, S.; Yamada, A. & Niki, S. (2009a). Efficiency enhancement of flexible CIGS solar cells using alkali-silicate glass thin layers as an alkali source material, *Conference Record of the Thirty-Forth IEEE Photovoltaic Specialists Conference 2009*, pp. 002349-002353, ISBN 978-1-4244-2949-3, Philadelphia, Pennsylvania, USA, June 7-12, 2009

Kapur, V. K.; Bansal, A. Phucan L. & Asensio, O. I. (2002). Non-vacuum printing process for CIGS solar cells on rigid and flexible substrates, *Conference Record of the Twenty-Ninth IEEE Photovoltaic Specialists Conference 2002*, pp. 688-691, ISBN 0-7803-7471-1, New Orleans, Louisiana, USA, May 19-24, 2002

Kessler, F.; Herrmann, D. & Powalla, M. (2005). Approaches to flexible CIGS thin-film solar cells. *Thin Solid Films*, Vol. 480-481, (December 2005), pp. 491-498, ISSN 0040-6090

Yagioka, T. & Nakada, T. (2009). Cd-Free Flexible Cu(In,Ga)Se₂ Thin Film Solar Cells with ZnS(O,OH) Buffer Layers on Ti Foils. *Applied Physics Express*, Vol. 2, No. 7, (June 2009), pp. 072201-1-072201-3, ISSN 1882-0778

Kapur, V. K.; Bansal, A. Le, P. & Asensio, O. I. (2003). Non-vacuum processing of CuIn₁₋ₓGaₓSe₂ solar cells on rigid and flexible substrates using nanoparticle precursor inks. *Thin Solid Films*, Vol. 431-432, (May 2003), pp 53-57, ISSN 0040-6090

Britt, J. S.; Wiedeman, S. Schoop, U. & Verebelyi, D. (2008). High-volume manufacturing of flexible and lightweight CIGS solar cells, *Conference Record of the Thirty-Thirdth IEEE*

Photovoltaic Specialists Conference 2008, pp. 574-577, ISBN 978-1-4244-2949-3, San Diego, California, USA, May 11-16, 2008

Gledhill, S.; Zykov, A. Allsop, N. Rissom, T. Schniebs, J. Kaufmann, C. A. Lux-Steiner, M. & Fischer, Ch-H. (2011). Spray pyrolysis of barrier layers for flexible thin film solar cells on steel. *Solar Energy Material and Solar Cells*, Vol. 95, (Febrary 2011) pp. 504-509, ISSN 0927-0248

Hashimoto, Y.; Satoh, T. Shimakawa, S. & Negami, T. (2003). High efficiency CIGS solar cell on flexible stainless steel, *Proceedings of third World Conference on Photovoltaic Energy Conversion 2003*, pp. 574-577, ISBN 4-9901816-0-3, Osaka, Japan, May 11-18, 2003

Khelifi, S.; Belghachi, A. Lauwaert, J. Decock, K. Wienke, J. Caballero, R. Kaufmann, C. A. & Burgelman, M. (2010). Characterization of flexible thin film CIGSe solar cells grown on different metallic foil substrates. *Energy Procedia*, Vol. 2, (August 2010), pp. 109-117, ISSN 1876-6102

Pinarbasi, M.; Aksu, S. Freitag, J. Boone, T. Zolla, H. Vasquez, J. Nayak, D. Lee, E. Wang, T. Abushama, J. & Metin, B. (2001). FLEXIBLE CELLS AND MODULES PRODUCED USING ROLL-TO-ROLL ELECTROPLATING APPROACH, *Conference Record of the Thirty-Fivth IEEE Photovoltaic Specialists Conference 2010*, pp. 000169-000174, ISBN 978-1-4244-5890-5, Honolulu, Hawaii, USA, June 20-25, 2010

Satoh, T.; Hashimoto, Y. Shimakawa, S. Hayashi, S. & Negami, T. (2000). CIGS solar cells on flexible stainless steel substrates, *Conference Record of the Twenty-Eighth IEEE Photovoltaic Specialists Conference 2000*, pp. 567-570, ISBN 0-7803-5772-8, Anchorage, Alaska, USA, September 15-22, 2000

Satoh, T.; Hashimoto, Y. Shimakawa, S. Hayashi, S. & Negami, T. (2003). Cu(In,Ga)Se$_2$ solar cells on stainless steel substrates covered with insulating layers. *Solar Energy Materials and Solar Cells*, Vol. 75, (January 2003), pp. 65-71, ISSN 0927-0248

Shi, C. Y.; Sun, Y. He, Q. Li, F. Y. & Zhao, J. C. (2009). Cu(In,Ga)Se$_2$ solar cells on stainless-steel substrates covered with ZnO diffusion barriers. *Solar Energy Materials and Solar Cells*, Vol. 93, (May 2009), pp. 654-656, ISSN 0927-0248

Wuerz, R.; Eicke, A. Frankenfeld, M. Kessler, F. Powalla, M. Rogin, P. & Yazdani-Assl, O. (2009). CIGS thin-film solar cells on steel substrates. *Thin Solid Films*, Vol. 517, (Febrary 2009), pp. 2415-2418, ISSN 0040-6090

Brémaud, D.; Rudmann, D. Kaelin, M. Ernits, K. Bilger, G. Döbeli, M. Zogg, H. & Tiwari, A. N. (2007). Flexible Cu(In,Ga)Se$_2$ on Al foils and the effects of Al during chemical bath deposition. *Thin Solid Films*, Vol. 515, (May 2007), pp. 5857-5861, ISSN 0040-6090

Ishizuka, S.; Yamada, A. Matsubara, K. Fons, P. Sakurai, K. & Niki, S. (2008a). Alkali incorporation control in Cu(In,Ga)Se$_2$ thin films using silicate thin layers and applications in enhancing flexible solar cell efficiency. *Applied Physics Letters*, Vol. 93, (September 2008), pp. 124105-1-124105-3, ISSN 0003-6951

Ishizuka, S.; Yamada, A. Fons, P. & Niki, S. (2008b). Flexible Cu(In,Ga)Se$_2$ solar cells fabricated using alkali-silicate glass thin layers as an alkali source material. *Journal of Renewable and Sustainable Energy*, Vol. 1, (Novenber 2008), pp. 013102-1-013102-8, ISSN 1941-7012

Ishizuka, S.; Yamada, A. Matsubara, K. Fons, P. Sakurai, K. & Niki, S. (2009b). Development of high-efficiency flexible Cu(In,Ga)Se$_2$ solar cells: A study of alkali doping effects on CIS, CIGS, and CGS using alkali-silicate glass thin layers. *Current Applied Physics*, Vol. 10, (November 2009), pp. S154-S156, ISSN 1567-1739

Ishizuka, S.; Yoshiyama, T. Mizukoshi, K. Yamada, A. & Niki, S. (2010). Monolithically integrated flexible Cu(In,Ga)Se$_2$ solar cell submodules. *Solar Energy Materials and Solar Cells*, Vol. 94, (July 2010), pp. 2052-2056, ISSN 0927-0248

Kapur, V. K.; Bansal, A. Le, P. Asensio, O. & Shigeoka, N. (2003). Non-vacuum processing of CIGS solar cells on flexible polymeric substrates, *Proceedings of third World Conference on Photovoltaic Energy Conversion 2003*, pp. 465-468, ISBN 4-9901816-0-3, Osaka, Japan, May 11-18, 2003

Brémaud, D.; Rudmann, D. Bilger, G. Zogg, H. & Tiwari, A. N. (2005). Towards the development of flexible CIGS solar cells on polymer films with efficiency exceeding 15%, *Conference Record of the Thirty-first IEEE Photovoltaic Specialists Conference 2005*, pp. 223-226, ISBN 0-7803-8707-4, Orlando, Florid, USA, January 3-7, 2005

Caballero, R.; Kaufmann, C. A. Eisenbarth, T. Unold, T. Schorr, S. Hesse, R. Klenk, R. & Schock, H.-W. (2009). The effect of NaF precursors on low temperature growth of CIGS thin film solar cells on polyimide substrates. *Physica Status Solidi A*, Vol. 206, (May 2009), pp. 1049-1053, ISSN 1862-6300

Ishizuka, S.; Hommoto, H. Kido, N. Hashimoto, K. Yamada, A. & Niki, S. (2008c). Efficiency Enhancement of Cu(In,Ga)Se$_2$ Solar Cells Fabricated on Flexible Polyimide Substrates using Alkali-Silicate Glass Thin Layers. *Applied Physics Express*, Vol. 1, No. 9, (September 2008), pp. 092303-1-092303-3, ISSN 1882-0778

Rudmann, D.; Brémaud, D. Zogg, H. & Tiwari, A. N. (2005). Na incorporation into Cu(In,Ga)Se$_2$ for high-efficiency flexible solar cells on polymer foils. *Journal of Applied Physics*, Vol. 97, (August 2005), pp. 084903-1-084903-5, ISSN 0021-8979

Zachmann, H.; Heinker, S. Braun, A. Mudryi, A. V. Gremenok, V. F. Ivaniukovich, A. V. & Yakushev, M. V. (2009). Characterisation of Cu(In,Ga)Se$_2$-based thin film solar cells on polyimide. *Thin Solid Films*, Vol. 517, (Febrary 2009), pp. 2209-2212, ISSN 0040-6090

Keyes, B. M.; Hasoon, F. Dippo, P. Balcioglu, A. & Abulfotuh, F. (1997). INFLUENCE OF Na ON THE ELECTRO-OPTICAL PROPERTIES OF Cu(In,Ga)Se$_2$, *Conference Record of the Twenty-Sixth IEEE Photovoltaic Specialists Conference*, pp. 479-482, ISBN 0-7803-3767-0, Anaheim, California, USA, September 29-October 3, 1997

Contreras, M. A.; Gabor, A. M. Tennant, A. Asher, S. Tuttle, J. & Noufi, R. (1994a). Accelerated publication 16.4% total-area conversion efficiency thin-film polycrystalline MgF$_2$/ZnO/CdS/Cu(In,Ga)Se$_2$/Mo solar cell. *Progress in Photovoltaics: Research and Applications*, Vol. 2, (October 1994) pp. 287-292, ISSN 1062-7995

Negami, T.; Satoh, T. Hashimoto, Y. Shimakawa, S. Hayashi, S. Muro, M. Inoue, H. & Kitagawa, M. (2002). Production technology for CIGS thin film solar cells. *Thin Solid Films*, Vol. 403-404, (January 2002), pp. 197-203, ISSN 0040-6090

Minemoto, T.; Okamoto, C. Omae, S. Murozono, M. Takakura, H. Hamakawa, Y. (2005). Fabrication of Spherical Silicon Solar Cells with Semi-Light-Concentration System. *Japanese Journal of Applied Physics*, Vol. 44, No. 7A, (July 2005), pp. 4820-4824, ISSN 0021-4922

Contreras, M. A.; Tuttle, J. Gabor, A. Tennant, A. Ramanathan, K. Asher, S. Franz, A. Keane, J. Wang, L. Scofield, J. & Noufi, R. (1994b). HIGH EFFICIENCY Cu(In,Ga)Se$_2$-

BASED SOLAR CELLS: PROCESSIMG OF NOVEL ABSORBER STRUCTURES, *Proceedings of the First World Conference on Photovoltaic Energy Conversion 1994*, pp. 68-75, ISBN 0-7803-1460-3, Waikoloa, Hawaii, USA, December 5-9, 1994

Minemoto, T.; Negami, T. Nishiwaki, S. Takakura, H. & Hamawaka, Y. (2000). Preparation of $Zn_{1-x}Mg_xO$ films by radio frequency magnetron sputtering. *Thin Solid Films*, Vol. 372, (August 2000), pp. 173-176, ISSN 0003-6951

Minemoto, T.; Hashimoto, Y. Satoh, T. Negami, T. Takakura, H. & Hamakawa, Y. (2001). $Cu(In,Ga)Se_2$ solar cells with controlled conduction band offset of window/ $Cu(In,Ga)Se_2$ layers. *Journal of Applied Physics*, Vol. 89, (June 2001), pp. 8327-8330, ISSN 0021-8979

9

Chemical Surface Deposition of CdS Ultra Thin Films from Aqueous Solutions

H. Il'chuk, P. Shapoval and V. Kusnezh
Lviv Polytechnic National University
Ukraine

1. Introduction

Solar cells (SC) are the most effective devices that allow direct one-stage conversion solar energy into electricity from the view of energy. The last yers tendency in traditional energetic forced to direct a significant part of research on the establishment of modern technology for production available and effective thin film SC that would not require the use of high temperature and pressure, a large number of rare and expensive materials. At the same time, to find ways for increase the conversion efficiency of solar energy it is necessary to understand the processes that occur in the elements. Therefore it is necessary to establish a correspondence between characteristics of elements and main structural, electronic and optical properties of initial semiconductor films. Therefore, the investigation of CdS thin films deposition process with desire photoelectric properties and fabrication on their base thin-film SC have great significance.

CdS is the main material for buffer layer in thin-film CdTe and Cu(In, Ga)Se$_2$ solar cells. It has a high photosensitivity and absorption, favorable energy band gap (Eg) 2,4 eV and photoconductivity (σ) 10^2 Om^{-1}cm^{-1} and does not change the properties with SC surface temperature increase during the work. One more peculiarity of this material is absence of the hole conduction due the acceptor additives and point defects recombination. Effective lifetime of the main carriers is very large (10...100 ms), that causes a initial photocurrent increase up to 10^5 times (Hamakawa, 2002). An important advantage of CdS thin films use in SC is possibility of their synthesis by different methods, including chemical deposition from solution which has significant preference over others: 1) grown nanocrystalites with a form close to spherical, while the electrochemical deposition - non-spherical (Jager-Waldau, 2004); 2) CdS thin films deposited from solution have structural, optical and electrical parameters thet do not inferior parameters of films received by other methods, but used equipment is available, simple, does not require use of the high temperatures and pressures compared, for example, with the vacuum evaporation or ion (sputtering or pulverization, spraying) methods; 3) the method is not explosive and low-toxic, compared with the vapor deposition methods; 4) enable control of the film growth and dynamically change the fabrication conditions for polycrystalline or smooth solid films.

2. Deposition of CdS thin films and structures based on

2.1 Fabrication methods

The thin film semiconductor properties largely depend on fabrication technology. Therefore development of actual methods, which would allow an influence on material parameters in the synthesis process and to obtain coating with the set properties, is an important scientific and technological problem. Recently the methods based on chemical processes dominate in the technology of metal sulfides thin films semiconductor. The semiconductor films with a thickness from several tenth of nanometers to hundreds of microns can be fabricated by a large number of so-called thin-film and thick-film methods. For large area in ground conditions aplication of thin-film solar cells crucial are not only their energy characteristics, but also their economic indicators. This causes use of bough thin film and thick-film technology methods for satisfying of such requirements as: fabrication simplicity, low cost, ability to create homogeneous films with a large area, controlling the deposition process, and ability to obtain the films with preferred structural, physical, chemical and electrooptical properties.

The deposition methods for wide range of semiconductors in detail are considered in literature (Aven & Prener, 1967, Chopra & Das, 1983, Green, 1998, Möller, 1993, Sze, 1981, Vossen & Kern, 1978,). We will consider only those methods that are used for cadmium sulfide films fabrication and are the best for solar cells producing. Thin film deposition process consists of three stages: 1) obtaining of substance in the form of atoms, molecules or ions; 2) transfer of these particles through an intermediate medium; 3) condensation of the particles on substrate. The methods of thin films fabrication are classified in several ways. Depending on the film grown phase are four methods of films deposition: 1) from the vapor phase; 2) from the liquid phase; 3) from the hydrothermal solutions; 4) from the solid phase. Depending on which way the vapour particle were obtained: using physical (thermal or ion sputtering), chemical or electrochemical processes, it is possible to classify deposition methods: physical vapor deposition; chemical vapor deposition; chemical deposition from the solution; electrochemical deposition. On the basis of physical and chemical vapor deposition were developed combined methods, such as: reactive evaporation, reactive ion sputtering and plasma deposition. Among the nonvacuum deposition methods of cadmium sulfide thin films for inexpensive solar cells with a large area perspective are: chemical deposition from baths (CBD), electrochemical deposition, mesh-screen printing, pyrolysis and pulverization followed by pyrolysis. Selection of the films deposition method first of all are specified by structural, mechanical and physical parameters, which should have thin-film sample.

Although, cadmium sulfide is the most widely studied thin film semiconductor material, interest of researchers to it is stable, and the number of scientific publications increasing all the time. Changing the deposition conditions drasticly alter electrical properties of CdS thin films. CdS films, obtained by vacuum evaporation have specific resistance $1 \cdot 10^3$ Om•cm and carrier concentration of 10^{16}-10^{18} cm^{-3}. Films always have n-type conductivity, that explains their structure deviation from stoichiometry, by sulfur vacancies and cadmium excess. Electrical properties of the films are largely depended from the concentration ratio of Cd and S atoms in the evaporation process and the presence of doping impurities. Electrical properties of CdS films, fabricated by pulverization followed by pyrolysis, are determined mainly by the peculiarities the chemisorption process of oxygen on grain boundaries, which accompanied by concentration decreaseing and charge carriers mobility. Due to presence of the sulfur vacancies such films always have n-type conductivity, and their resistance can vary widely, differing by the amount of eighth order. Epitaxial CdS films are characteristic due to carrier high mobility. With the increase of substrate temperature concentration of carriers grows by an exponential law. This increase the electron mobility. Optical properties of CdS films are strongly dependent on their microstructure and thus on the method and conditions of deposition. For example, evaporation of CdS results in smooth mirror reflective films, but increasing their thickness leads to a predominance of diffuse reflection. The CdS films, obtained by ion sputtering have the area with rapid change of transmission at 520 nm, corresponding CdS band gap. In the same time in the long-wave spectral range films have high transparency.

2.2 Use of the CdS films in photovoltaic cells

Edmund Becquerel, a French experimental physicist, discovered the photo-voltaic efect in 1839 while experimenting with an electrolytic cell, made up of two metal electrodes placed in an electricity-conducting solution. He observed that current increased when the electrolytic cell was exposed to light (Becquerel, 1839). Then in 1873 Willoughby Smith discovered the photoconductivity of selenium. The first selenium cell was made in 1877 (Adams, 1877), and five years later Fritts (Fahrenbruch & Bube, 1983) described the first solar cell made from selenium wafers. By 1914 solar conversion eficiencies of about 1 % were achieved with the selenium cell after it was finally realized that an energy barrier was involved both in this cell and in the copper/copper oxide cell.

The modern era of photovoltaics started in 1954. In that year was reported a solar conversion efficiency of 6 % (Chapin at al., 1954) for a silicon single-crystal cell. In 1955 Western Electric began to sell commercial licenses for silicon PV technologies. Already in 1958 silicon cell efficiency under terrestrial sunlight had reached 14 %. At present, available in the market SC are mainly represented of monocrystalline silicon SC. Through high-temperature process of their formation, crystal (from ingots grown from melt by Czochralski method) and polycrystalline silicon solar cells have too high price, to be seen as a significant competitor to the formation of energy from solid fuels. Polycrystalline silicon provides lower expenses and increase production, rather than crystalline silicon. In 1998, approximately 30 % photovoltaic world production was based on the polycrystalline silicon wafers. Nowadays solar cells conversion efficiency based on monocrystalline silicon is 25 %, polycrystalline – 20 % (Green at al., 2011).

In 1954 reported 6 % solar conversion efficiency (Reynolds at al., 1954) in what later came to be understood as the cuprous sulfide/cadmium sulfide heterojunction (HJ). This was the first all-thin-film photovoltaic system to receive significant attention. In following years the efficiency of Cu_xS/CdS increased up to 10 % and a number of pilot production plants were installed, but after several years of research it was realized that these solar cells have unsolvable problems of stability owing to the diffusion of copper from Cu_xS to CdS layers. By taking advantage of new technology, work out on Cu_xS/CdS, researchers have rapidly raised the effciency of the gallium arsenide based cell with 4 % efficiency (Jenny at al., 1956) to present eficiencies exceeding 27 % (Green at al., 2011).

However in the last 20 years other thin films solar cells have taken the place of the cuprous sulfide/cadmium sulfide, and their eficiency have raised up to almost 20 %. The most predominant are two: copper indium gallium diselenide/cadmium sulfide $(Cu(In,Ga)Se_2/CdS)$ and cadmium telluride/cadmium sulfide (CdTe/CdS). The first CdTe heterojunctions were constructed from a thin film of n-type CdTe material and a layer of p-type copper telluride ($Cu_{2-x}Te$), producing ~7 % eficient CdTe-based thin-film solar cell (Basol, 1990). However, these devices showed stability problems similar to those encountered with the analogous $Cu_{2-x}S/CdS$ solar cell, as a result of the difuusion of copper from the p layer. The lack of suitable materials with which to form heterojunctions on n-type CdTe, and the stability problems of the $Cu_{2-x}S/CdS$ device, stimulated investigations into p-CdTe/n-CdS junctions since the early 1970s. Adirovich (Adirovich at al., 1969) first deposited these films on TCO-coated glass; this is now used almost universally for CdTe/CdS cells, and is referred to as the superstrate configuration. In 1972 5-6 % efficiencies were reported (Bonnet & Rabenhorst, 1972) for a graded band gap CdS_xTe_{1-x} solar cell.

The research for $CuInSe_2/CdS$ started in the seventies, a 12 % efficiency single-crystal heterojunction p-$CuInSe_2$/n-CdS cells were made by in 1974 (Wagner, 1975) and in 1976 was presented the first thin film solar cells with 4-5 % eficiency (Kazmerski at al., 1976). In the last 30 years a big development of these cells was given by the National Renewable Energy Laboratories (NREL) in U.S.A. and by the EuroCIS consortium in Europe.

Nowadys CdS among Si, Ge, CdTe, Cu(In, Ga)Se$_2$, ZnO belongs to the widespread group of semiconductors. Beyond the attention of researchers are still many issues associated with cadmium sulfide as componenet of thin-film semiconductor devices, although the CdS is one of the most studied semiconductor materials.

2.3 Peculiarities of chemical bath deposition (CBD)

CBD technology consist of the deposition of semiconductor films on a substrate immersed in solution containing metal ions and hydroxide, sulfide or selenide ions source. The first work on CBD is dated 1910 and concerns to the PbS thin films deposition (Houser & Beisalski, 1910). Basic principles underlying the CBD of semiconductor films and earlier studies in this field were presented in the review article (Hass at al., 1982), which encouraged many researchers to begin work in this direction. Further progress in this area is presented in review article (Lokhande, 1991), where references are given for 35 compounds produced by the mentioned method, and other related links. Chemical reactions and CBD details for many compounds were listed in the next paper (Grozdanov, 1994). The number of materials which can be produce by CBD, greatly increased, partly due to the possibility of producing multilayer film structures by this method with subsequent annealing, which stimulates crosboundary diffusion of metal ions and thereby motivates fabrication of new materials with high thermal stability. For example, crossboundary diffusion of CBD coatings PbS/CuS and ZnS/CuS leads to materials such as $Pb_xCu_yS_z$ and $Zn_xCu_yS_z$ with p-type conductivity and thermal stability up to 573 K (Huang at al., 1994). Annealing of Bi_2S_3/CuS coatings at temperatures 523-573 K leads to formation of new Cu_3BiS_3 compounds with p-type conductivity (Nair at al., 1997). In recent years we counted approximately 120 CBD semiconductor compouns.

Among the first applications of CBD semiconductor films were photodetectors based on PbS and PbSe (Bode at al., 1996). Although the chemically precipitated CdS films were made back in the 60's of last century, for photodetectors were used CdS layers, obtained by screen printing and sintering (Wolf, 1975). Chemically deposited CdSe films are fully suitable for use in photodetectors (Svechnikov & Kaganovich, 1980). At late 70's and early 80-ies the main direction in chemical bath deposition technology was deposition of thin films for use in solar energy conversion. One of the first developments in this area was the coating producing that absorbs sunlight (Reddy at al., 1987), and its use in glass vacuum tube collectors (Estrada-Gasca at al., 1992). Application of the chemically deposited films in coatings for controlling the flow of sunlight was first proposed in 1989 (Nair at al., 1989). The efficiency improving of such coatings in glass vacuum tube collectors were presented in (Estrada-Gasca at al., 1993). One of the main applications of chemically deposited semiconductor films has been their use in photoelectrochemical SC, mostly CdS and CdSe films (Hass at al., 1982, Boudreau & Rauh, 1983, Rincon at al., 1998). The use of chemically deposited semiconductor films in thin SC has a short history. In the structure Mo/CuInSe$_2$/CdS/ZnO, which showed 11% efficiency (Basol & Kapur, 1990), was by the first time used chemically deposited CdS thin film. Further structure improvement allowed to reach 17% efficiency (Tuttle at al., 1995). Chemically deposited CdS film with thickness of 50 nm has been an essential element of this structure. The biggest, confirmed today for SC based on CdS/CdTe, is 16,5% efficiency in which CdS film was chemically deposited in bath (Green at al., 2011). Entering highly resistive CdS film in p-CuInSe$_2$/CdS/n-CdS solar cell structure deemed necessary step towards improving of the solar cells stability (Mickelsen & Chen, 1980). Performed theoretical calculations (Rothwarf, 1982) showed that the thickness of CBD CdS films should be as small as possible to increase efficiency of solar cells with its use. Therefore, chemical deposition technology, which allows to fully cover the substrate at small film thickness was selected for the fabrication of thin films and showed significantly better results (Basol at al., 1991). Efficiency of n-CdSe or n-Sb$_2$S$_3$ chemically deposited films with WO$_3$ inclusions as absorber material in solar cells based on the Schottky barrier has been proved in practice. For example, elements on the Schottky barrier ITO/n-CdSe(5 µm)/Pt/Ni/Au (13 nm) shows U_{xx}=0,72 V, I_{k3}=14,1 mA cm^{-2}, fill factor 0,7, and 5,5% efficiency (Savadogo & Mandal, 1993 & 1994). Abovementioned possible applications of chemical bath deposition, particularly in solar energy conversion, provided the growing interest to chemical deposition of semiconductor thin films. Chemical deposition is perfect for producing thin films on large areas and at low temperatures, which is one of the main requirements for the mass use of solar energy.

2.4 The advantages of chemical surface deposition (CSD) over CBD

In the CBD process, the heat necessary to activate chemical reaction is transferred from the bath to the sample surface, inducing a heterogeneous growth of CdS on the surface and homogeneous CdS formation in the bath volume. The reaction is better in the hottest region of the bath. Therefore, for baths heated with thermal cover deposition also occurs on the walls, and bath, which heat up immersed heater, significant deposition occurs on heating element. Additionally, the solution in the bath should be actively mixed to ensure uniform thermal and chemical homogeneity and to minimize adhesion of homogeneously produced particles to the surface of CdS film. The disproportion of bath volume and that which is necessary for the formation of CdS film, leads to significant proportion of wastes with high cadmium content. Different groups of researchers put efforts for decreasing the ratio of volumes bath/surface through use of overlays. However the clear way for unification of large areas deposition with high cadmium utilization and high speed of growth, to achieve high efficiency of transformation is not represented.

The chemical surface deposition (CSD) technology demonstrated in this paper overcomes these limitations through use of the sample surface as a heat source and use of solution surface tension to minimize the liquid volume. The combination of heat delivery method to surface and small volume of solution leads to high utilization of cadmium and its compounds.

This paper describes CSD technology of CdS thin films from aqueous solutions of cadmium salts $CdSO_4$, $CdCl_2$, CdI_2. The properties of CdS films deposited on glass and ITO/glass from the nature of the initial salt and solar cells based on CdTe/CdS with CSD CdS films as windows was investigated.

3. Chemical surface deposition of CdS thin films from $CdSO_4$, $CdCl_2$, CdI_2 aqueus solutions

3.1 Introduction

One of the methods to increase SC efficiency based on CdS/CdTe, $CdS/CuIn_{1-x}Ga_xSe_2$, with the CdS film as the window is increasing the current density value (Stevenson, 2008). This can be achieved by reducing losses in the photons optical absorption from $\lambda > 500$ nm by reducing CdS film thickness. To provide a spatially homogeneous work of the device the CdS films should not only be thin, but solid, durable and resistant to further technology of SC production. To produce ultra-thin (from 30 to 100 nm) and homogeneous CdS films the technology of bath chemical deposition is widely used (Estela Calixto at al., 2008, Mugdur at al., 2007).

Chemical deposition technology is quite simple, inexpensive and suitable for the deposition of polycrystalline CdS films on large areas. Deposition of thin CdS films from aqueous solutions is a reaction between cadmium salt and thiocarbamid (thiourea) in alkaline medium. Mostly are used simple cadmium salts: $CdSO_4$ (Chaisitsak at al., 2002, Contreras at al., 2002, Tiwari & Tiwari, 2006, Chen at al., 2008), CdI_2 (Nakada & Kunioka, 1999, Hashimoto at al., 1998), $Cd(CH3COO)_2$ (Granath at al., 2000, Rau & Scmidt, 2001) and $CdCl_2$ (Qiu at al., 1997, Aguilar-Hernández at al., 2006). Thiourea (TM) is used as sulfide agent in the reactions of sulfide deposition, as has a high affinity to metal cations and decomposes at low temperatures. Deposition process can be described by two mechanisms (Oladeji, 1997, Soubane, 2007). Homogeneous mechanism involves formation of layer with the CdS colloidal particles, which are formed in solution and consists of several stages.

1. Ammonium dissociation:

$$NH_4^+ + OH^- \rightleftarrows NH_3 + H_2O \tag{1}$$

In alkaline medium due to interaction Cd^{2+} ions with the OH- environment ions is possible formation of undesirable product - $Cd(OH)_2$:

$$Cd^{2+} + OH^- \rightarrow Cd(OH)_2 \downarrow \tag{2}$$

2. Thiourea hydrolysis $(NH_2)_2CS$ with the the formation of sulfide ions

$$(NH_2)_2CS + H_2O \rightleftarrows HS^- + H^+ + (NH_2)_2CO \tag{3}$$

$$HS^- + OH^- \rightleftarrows S^{2-} + H_2O \tag{4}$$

3. Final product formation

$$Cd^{2+} + S^{2-} \rightleftarrows CdS \downarrow \tag{5}$$

Deposition of thin CdS films from the aqueous solutions through the stage of cadmium tetramin $[Cd(NH_3)_4]^{2+}$ complex ion formation, which reduces the overall speed of reaction and prevents $Cd(OH)_2$ formation by the heterogeneous mechanism.

$$Cd^{2+} + 4NH_4OH \rightarrow [Cd(NH_3)_4]^{2+} + 4H_2O \qquad (6)$$

$$[Cd(NH_3)_4]^{2+} + S^{2-} \rightarrow CdS \downarrow + 4NH_3 \qquad (7)$$

In general form:

$$[Cd(NH_3)_4]^{2+} + (NH_2)_2CS + OH^- \rightarrow CdS \downarrow + 4NH_3 + H^+ + (NH_2)_2CO \qquad (8)$$

The sulphides films deposition from thiocarbamid coordination compounds has some chemical peculiarities. Depending on the nature and the salt solution composition may be dominated different coordination forms, and with thiourea molecules in complex inner sphere may contain anions Cl⁻, Br⁻, J⁻, and $SO_4{}^{2-}$ under certain conditions. Thus, the cadmium atoms close environment are atoms of sulfur, halogens and oxygen, and at the thermal decomposition part of the Cd-Hal or Cd-O bonds are stored and in the sulfide lattice are formed Hal$_S$• and O$_S$•defects. In conjunction with the substrate the thiocarbamid complexes orientation on active centers of its surface is observed. The complex particles that can interact with active centers on the substrate are the link that provides sulfide link with the substrate. The nature of this interaction determines the nature of film adhesion. In the case of cadmium sulfide deposition on quartz or glass substrates the active centers are sylanolane groups (≡Si–OH) which interact with halide or mixed hydroxide complexes. In result of such interaction the Cd–O–Si oxygen bridges are created. This explains the good adhesion of the cadmium sulfide films deposited from thiocarbamid coordination compounds to glass substrates (Palatnik & Sorokin, 1978).

3.2 Chemical surfact deposition of CdS thin films

In CSD, a solution at ambient temperature containing the desired reactants is applied to a pretreated surface. Glass or ITO/glass (16×20 mm) substrates, CdTe (10×10 mm) and Si (30×20 mm) wafers were used in the entire work. After that sample with working solution is heated and endured for a given temperature (Fig. 1). To ensure uniformity of heating plate

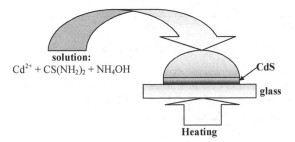

$$[Cd(NH_3)_4]^{2+} + CS(NH_2)_2 + OH^- \leftrightarrow CdS\downarrow + 4N_3 + H^+ + (NH_2)CO$$

Fig. 1. Scheme of CdS films thin chemical surface deposition

with working solution is previously placed on thermostated (343 K) surface. Surface tension of the solution provides a minimum volume of reaction mixture and its maintenance on the substrate. Film deposition occurs through the heterogeneous growth of compounds on the substrate surface by transfer of heat to the work solution. Heterogeneous growth is preferred over homogeneous loss due to thermal stimulation of chemical activity on warmer surface. At a result we receive a high proportion of cadmium from a solution in film and depending on the substrate, the heteroepitaxial film growth. The outflow of heat from the solution to environment helps to keep the favorable conditions for the film heterogeneous growth in time required for film deposition. After heating the plate was removed, the surface was rinsed with distilled water and dried in the air.

The combination of factors of the heat delivery to phase division surface (substrate-solution) and small volume of working solution in the CSD allows to receive coverage with satisfactory performance, increase the efficiency of the reagents, and therefore simplify their utilization. For deposition of CdS films were used freshlyprepared aqueous solutions of one of three cadmium salts: $CdSO_4$, $CdCl_2$, CdI_2. Solution ingredients and the corresponding concentrations are presented in Table. 1.

salt	C(cadmium salt), mol/l	$C(CS(NH_2)_2)$, mol/l	$C(NH_4OH)$, mol/l
$CdSO_4$			
$CdCl_2$	0,001; 0,0001	0,1; 0,01	1,8; 1,2
CdI_2			

Table 1. Ingredients and concentrations of solutions for CSD of CdS films, T=343 K, pH=12

Several modifications of films CSD were used. First modification (A) includes single applying of working solution and it different time exposure (5 to 12 min.) on the substrate. The second modification (B) provided repeated addition (3 min intervals.) of fresh working solution on the substrate surface. The difference of the third modification (C) consistent in applying (with 3 min. time exposure) and subsequent flushing of working solution on the substrate surface, ie in layer deposition. In such way we achieved increase and regulation of CdS film thickness.

	modifications		
	A	B	C
maximum thickness, nm	62	65	105
deposition rate, nm/min.	≤6	4-6	≥8

Table 2. The CdS films maximum thickness and deposition rate depending on the CSD modification

Aplying of A modification results in the smallest CdS film thickness, as seen from Table. 2. This is because the main part of the film (80-90 % thickness) is deposited in 2-3 min. Further time exposure of the working solution-substrate system is not accompanied by visible changes in the appearance of the formed film, apparently due to exhaustion of working solution. Therefore, during the multistage (CSD modifications B and C) CdS films deposition the duration of elementary expositions deposition was 3 min. Based on the structural studies results for further work modification B was selected.

3.3 Properties of CSD CdS thin films

The film thickness was determined by ellipsometric measurement of light polarization change after light reflection from an air-film interface on the LEF-3M instrument, allowing precision from 5 to 10 nm, for film thickness less then 100 nm. Morphology of the film surface and the elemental composition were investigated using the scanning electron microscopes REMMA-102-02 with EDS and WEDS and JSM-6490LV. Crystallinity of the CdS film structure was investigated using the automated X-ray diffractometer HZG-4A (with CuK_{α} radiation, λ=0,15406 nm). The optical transmission measurements have been done at room temperature with unpolarized light at normal incidence in the wavelength range from 300 to 1000 nm using Shimadzu UV-3600 double beam UV/VIS spectrophotometer. The optical absorption coefficient α was calculated for each film using the equation

$$I_t = I_0 \exp(-\alpha t) \tag{9}$$

where t is the film thickness, I_t and I_0 are the intensity of transmitted light and initial light, respectively. The absorption coefficient α is related to the incident photon energy $h\nu$ as:

$$\alpha \cdot h\nu = A\left(h\nu - E_g\right)^{n/2} \tag{10}$$

where A is a constant dependent on electron and hole effective mass and interband transition, E_g is the optical band gap, and n is equal to 1 for direct band gap material such as CdS. The band gap E_g was determined for each film by plotting $(\alpha h\nu)^2$ vs $h\nu$ and then extrapolating the straight line to the energy axis.

3.3.1 Thickness and deposition rate

The peculiarity of the CSD method is that after the first deposition the function of the substrate is performed not by glass, but by formed CdS film. All subsequent depositions are conducted on the same substrate. Through this growth rate of successive layers is approximately the same, and the total film thickness increases in equal size. The data of film thickness measurements and calculated average growth rate is shown on Fig. 2. The accuracy of ellipsometric measurements of thickness increased as the total thickness of the film growth, so that the absolute error varied from ± 10 nm to ± 5 nm. The highest thickness obtained was in the case of $CdSO_4$, and the least thickness in the CdI_2 case.

Apparently, among all other Cd salts, CdI_2 always results in a much thinner film. This observation was in agreement with what was reported earlier (Kitaev at al., 1965, Ortega-Borges & Lincot, 1993). This can be explained by different values of stability constant of Cd complexes complementary (Khallaf at al., 2008).While using for CSD the $CdCl_2$ (Fig. 2, a) were obtain almost linear dependence increase of film thickness on the deposition time. For films deposited with $CdSO_4$ and CdJ_2 (Fig. 2, b and c, respectively), the dependence of film thickness on deposition time was more complicated, but also had a character close to linear. This fact can be used for CdS films thickness control with high precision in the CdS/SdTe HJ fabrication. The differences in the nature of layer growth of thin CdS films can be explained by the process stages. When solution is applied to the substrate and heated, thiocarbamid complexes start to orient on active centers of the substrate surface and form CdS growth centers. The maximum possible number of growth centers is determined by the number of active centers on the substrate surface, which is considerably less than reactive particles in solution. Under the influence of continuous solution flow the grow centers increases and turn into islands. After a surface filling the islands are merging and form netted

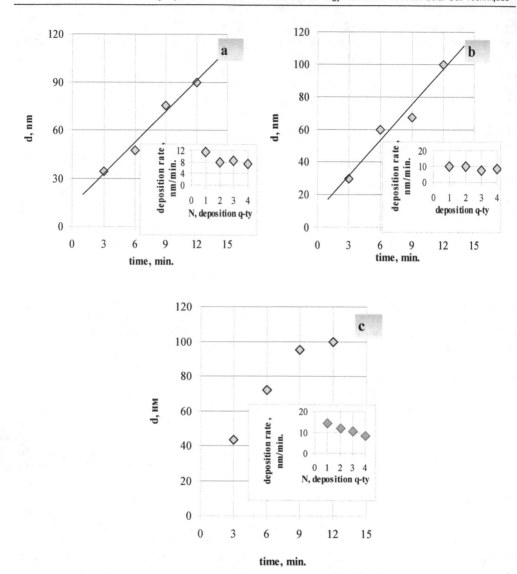

Fig. 2. The CdS thin film thickness dependence on time and quantity of deposition from aqueus solution: CdCl$_2$ (a); CdSO$_4$ (b); CdJ$_2$ (c). The mean deposition rate of CdS thin films on figure inset.

structures that consist of pores and channels. Further film growth is, in fact, filling the pores and channels. It slows the increase of film thickness, but does not alter the film weight gain. At later stages of the growth occurs reflection of the particles stream from the surface that leads to film growth rate decrease, and in the future - to its almost complete stop. Maximum growth rate of CSD at 343 K had films deposited from CdI$_2$ solution. Big deposition rates cause to the significant film defections, which confirm the results of their structural studies.

3.3.2 Surface morphology

The results of the CdS films investigation by scanning electron microscopy, deposited from diferent aqueous salt solutions are shown in Fig. 3-7, in the reflected and secondary electrons mode.

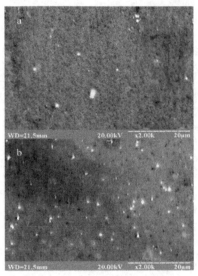

Fig. 3. Surface morphology of CdS film deposited from $CdSO_4$ aqueus solution, A modification (a) and C modification (b). REMMA-102-02, accelerating voltage 20 kV, scale 1:2000

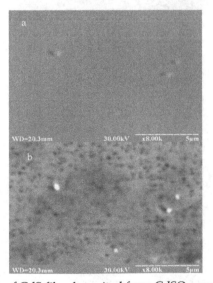

Fig. 4. Surface morphology of CdS film deposited from $CdSO_4$ aqueus solution on ITO coated glass in the secondary-electron mode (a) and reflected-electron mode (b). REMMA-102-02, accelerating voltage 30 kV, scale 1:8000

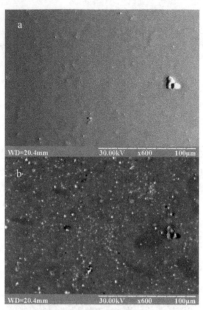

Fig. 5. Surface morphology of CdS film deposited from $CdCl_2$ aqueus solution in the secondary-electron mode (a) and reflected-electron mode (b). REMMA-102-02, accelerating voltage 30 kV, scale 1:600

Fig. 6. Surface morphology of CdS film deposited from CdI_2 aqueus solution in the secondary-electron mode (a) and reflected-electron mode (b). REMMA-102-02, accelerating voltage 30 kV, scale 1:1200

In reflected electron mode the photo qualitatively displays the surface composition (the lighter point, the heavier elements), and in secondary electron mode - the surface morfology. As seen all CdS film fabricated by C modification completely covers the substrate across the sample area, are homogeneous and solid. In reflected electrons mode are observed white dots indicating the localization heavier compared to the film phase. Comparison of CdS film images, obtained in both reflected and secondary electrons (Fig. 4-6), indicates that the heavier phase inclusions are on the film surface.

So, these heavier phase inclusions are particles on the surface (surface macrodefects) and most likely were formed in the final phase of deposition. The concentration of macrodefects on the surface in the investigated CdS films deposited from varus cadmium salts are presented in table 3. Regardles of applied salt surface macrodefects concentration is almost the same and is 100 times smaler than for CBD films (Romeo at al., 2003). Using EDS and WDS measurements, the stoichiometry of all films were studied. The generalized results of the surface morphology and X-ray microanalysis investigation of thin CdS films, deposited from various cadmium salts are given in Table 3. We determined that the particles on the CdS films surface (macrodefects) are CdS particles with a different stoichiometry than the film. The stoichiometry deviation towards sulfur is quite unexpected because in most nonvacuum deposition methods the lack of sulfur is observed.

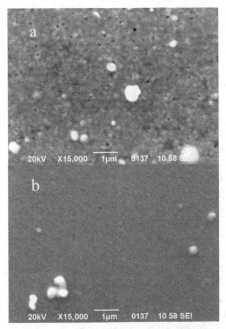

Fig. 7. Surface morphology of CdS film deposited from CdI_2 aqueus solution before (a) and after annealing (b). JSM-6490LV, accelerating voltage 20 kV, scale 1:15000

The CdI_2-based films had composition close to stoichiometric while the $CdSO_4$-based films showed the biggest deviation from stoichiometric composition that agre with results of CBD (Ortega-Borges & Lincot, 1993) (Table 3). Sulfur excess in CSD CdS films gives us the opportunity to perform annealing in the normal (air), not sulfur medium because they do not need to enter in film extra amount of sulfur to ensure stoichiometry. Analysis of CdS films surface morphology, obtained by AFM (Fig. 8) shows that the method of deposition and the nature of the initial cadmium-containing salt have significant affect on the CdS film surface structures. Using the deposition B modification ensure much more evenly cover over the sample area than A modification. The best results were obtained by C modification. The CdS films deposited from CdSO$_4$ aqueous solution by B and C (Fig. 8, a and b, respectively) have different surface morphology. The surface of all films obtained in the C modifications, is completely packed with crystalline grains. The exception is the film obtained from cadmium iodide aqueous solution. Along with the films surface morphology the results of surface roughness analysis are presented.

salt	surface macrodefects concentration, cm^{-2}	Cd/S rate on film surface	Cd/S rate of surface macrodefects
CdSO$_4$	10^6–10^7	0,880	0,800
CdCl$_2$	10^7	0,898	0,908
CdI$_2$	10^6–10^7, the pineholes are observed, for films deposited from two other salts the pinholes are almost absent	0,911	1,061

Table 3. Summarized results of surface morphology and X-ray microanalysis investigation of CdS thin films, deposited from various cadmium salts

3.3.3 Crystal structure
Experimental diffraction intensities of CdS films, obtained by B and C modification of (curves 2 and 3), respectively, are shown in Fig. 9. In all tested samples polycrystallinity of CdS films is expressed with the noticeable presence of cubic phase. The curves 2 and 4 on fig. 9. indicates that the samples are almost completely polycrystalline.
The first 26,45^0 peak of cubic phase (curves 2 and 4) is slightly expressed and shifted compered to the corresponding XRD peak from single CdS crystal (curve 1), which can be explained by the small size of grains as the probability of mechanical stress in films is very small because of low speed growth (Table 2).
In addition to the 26,45^0 peak on curve 3 (Fig. 9) are present two more – 43,90^0 and 52,00^0, corresponding to the cubic phase. Implemented sample heat treatment does not result in a significant increase in the intensity of any of the three peaks, and even the intensity of first one decreases (curve 4). The shift of the first (26,45^0) peak (curves 3 and 4) related with a decrease after annealing of mechanical tensions in the film, and intensity decrease of this peak indicates a simultaneous transition in polycrystalline cubic phase. Size grains expected increase by recrystallization has not occurred. Thus, annealing conditions to improve crystallinity of films need correction. Based on the data diffraction pattern most of the cubic phase is contained in the films deposited by C modification C(CdSO$_4$) = 0,001 mol/l. The transition to the hexagonal phase after annealing is not observed, unlike CBD CdS film (Archbold at al., 2005, Romeo at al., 2000).
Fig. 10 shows the experimental diffraction intensities obtained from CdS films, deposited from aqueous solutions of CdSO$_4$, CdCl$_2$, CdI$_2$ salts on glass substrates before and after annealing. In all tested samples polycrystallinity of CdS films is expressed with the noticeable presence of cubic phase. From the curves 2 (Fig. 10, a, b, c) can be seen that as deposited samples are almost entirely polycrystalline. The first peak of 26,45 ° for the cubic

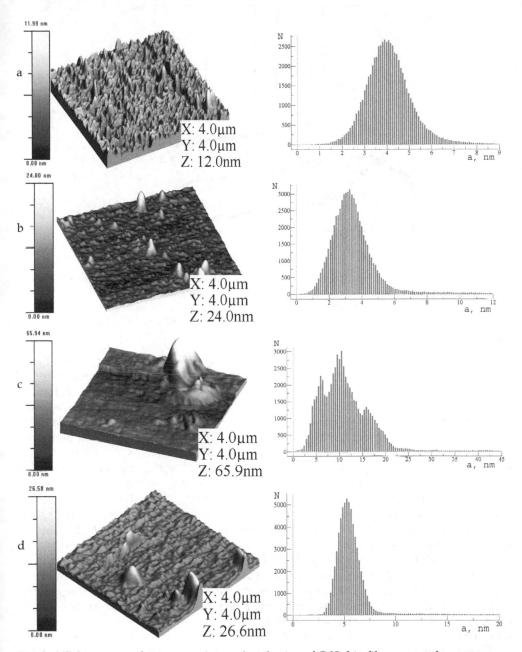

Fig. 8. AFM images and mean roughness distribution of CdS thin films grown from aqueus solution: CdSO$_4$, B modification (a); CdSO$_4$, C modification (b); CdCl$_2$, C modification (c); CdJ$_2$, C modification (d)

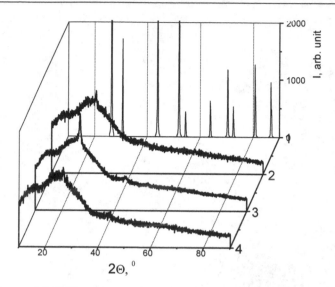

Fig. 9. XRD pattern of CdS film deposited on glass substrate from $CdSO_4$ aqueous solution with $C(CdSO_4)$=0,001 mol/l by B and C modification (curves 2 and 3); $C(CdSO_4)$=0,001 mol/l by C modification, after annealing (curve 4); with $C(CdSO_4)$=0,0001 mol/l C modification (curve 5); CdS cubic monocrystal reference pattern (curve 1).

phase have low intensitivity and is slightly shifted against the corresponding peak of CdS single crystal. This can be explained by the small size of grains as the probability of mechanical tensions in films deposited from $CdSO_4$, $CdCl_2$ salts solutions is neglectible due too low growth speed. Besides peak 26,45°, on curve 2 (Fig. 10 b) are present two more - 43,90° and 52,00° corresponding to the cubic phase. The heat treatment of samples leads to a significant increase in the intensity of the first two peaks for films deposited from $CdSO_4$, CdI_2. For films deposited from $CdCl_2$ aqueous solutions, the nature of XRD curve practically unchanges due to annealing. For CdS films, deposited from CdI_2 aqueous solution (fig. 10c curve 1) after annealing were observed intensity increases of 26,45°, 52,00° peaks and the appearance of third peak 43,90°. This indicates a reduction of disordered polycrystalline phase which transforms into crystalline phase and a rather significant restructure of CSD film, that cinsides with the results of the surface morphology investigations (Fig. 7).

Experimental diffraction intensities of CdS films deposited on Si and CdTe substrates are presented in Fig. 11. As seen, the results for various substrates were different, but both express polycrystallinity of CdS films. Besides the peaks of Si (Fig. 11 a, № 3) and CdTe (Fig. 11 b, № 5, 8) substrates are present a significant number of peaks corresponding to different phases of CdS compound. These results indicate the existence of a mixture of two structural phases (cubic and hexagonal) that is often observed for CdS films fabricated by nonvacuum methods (Calixto & Sebastian, 1999). The X-ray diffraction peaks N 1, 2, 7, 4, 9 (Fig. 11a) on silicon corespond to hexagonal structure, cubic may respond only the peak number 1. For films on CdTe substrate peaks intensity of hexagonal and cubic phases is much higher.

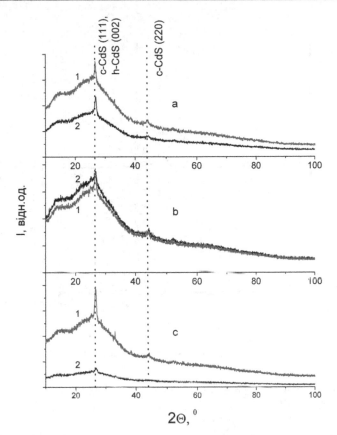

Fig. 10. XRD pattern of CdS film deposited on glass substrate from aqueus solution: CdSO₄ (a); CdCl₂ (b); CdJ₂ (c); as deposited (2); after annealing (1).

3.3.4 Optical properties

Absorption coefficient in the fundamental absorption area for all CdS samples was 10^5 cm^{-1}. The absorption spectra of samples (Fig. 12.) clearly shows the existence of the CdS compounds in all films deposited from aqueous solutions of cadmium-containing salts. Spectral dependence of CdS films absorption in the coordinates $(\alpha^* h\nu)^2$ vs $h\nu$ demonstrate the presence of fundamental absorption edge (Fig. 12), localized in the region 2,5 eV. The calculated band gaps of the films are in good agreement with the reported values (Landolt-Börnstein, 1999, Aven & Prener, 1967) and correspond to the direct allowed band transition. We do not observe a straight-line behaviour on graphs of $(\alpha h\nu)^{2/3}$ vs $h\nu$ (direct forbidden), $(\alpha h\nu)^{1/2}$ vs $h\nu$ (indirect allowed) $(\alpha h\nu)^{1/3}$ vs $h\nu$ (indirect forbidden). These plots (not shown) reveal that the type of transition is neither direct forbidden nor indirect. For films deposited in the same technological modes on glass and ITO coated glass substrates, the location of fundamental absorption edge are olmost the same. A small (0,01 eV) difference between the fundamental absorption edge values for films on glass and ITO/glass are caused by the difference of substrates surface roughness.

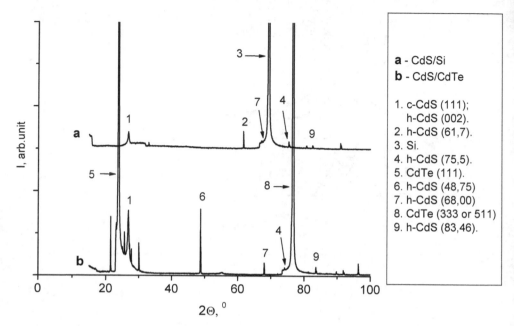

Fig. 11. XRD pattern of CdS films deposited from $CdCl_2$ aqueus solution on Si (a) and CdTe (b) substrates

The fundamental absorption edge localization feature in CdS films, in comparison with CdS monocrystal, is that in films it is shifted to higher energy region (2,537 eV and 2,547 eV for films on glass and ITO/glass, respectively).

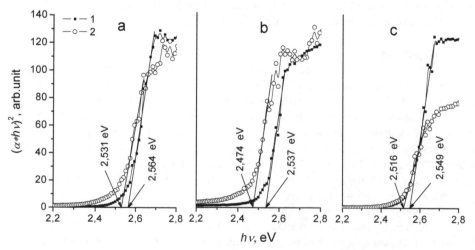

Fig. 12. Optical absorbance spectra of CdS film deposited on glass substrate from aqueus solution: $CdSO_4$ (a); $CdCl_2$ (b); CdJ_2 (c); as deposited (1); after annealing (2).

This allows to expand CdS/CdTe solar cells phototransformation area and increase their efficiency. Reducing energy fundamental absorption edge of CdS films after annealing (Fig. 12, curves 2) can be coused by grain growth (Nair at al., 2001). Sharpest edge of fundamental absorption have CdS films, deposited on glass substrate. This indicates a smaller number of macro defects in these films compared with annealed. Energy levels of this defects are lying near the edge zones. The increase long-wave "tail" of the absorption curve for annealed film (Fig. 12, a, b, curve 2) is caused by increase of absorption near the CdS film surface, where in the process of annealing in air CdO can be formed.

All films have high transmission, with the transmission in the CdI$_2$ case being better than that of the other three films. This was expected, since the SEM micrographs, showed the pinholes on CdI$_2$-based film. The lowered transmission of our films is caused by their surface roughness, due to coverage by surface macrodefects which are overgrowth crystallite, causes light scattering. Spectral dependence of optical transmission in the visible region of CSD CdS films before and after annealing are shown in Fig. 13. The main feature of the annealed CdS films spectra is small (0.033 eV) shift of fundamental absorption edge in the longwave region and reducing the optical transmission more than 20%. Reducing the transmission is determined not only by absorption and reflection from the film surface, but olso by quite significant changes in the film structure after annealing. From the SEM holes were observed (Fig. 7) after annealing they completely disappeared. X-ray pattern (Fig. 10, c, curve 1) also confirm a significant increase in the film's crystallinity structure as a result of annealing, despite indifferent directing effect of the glass substrate.

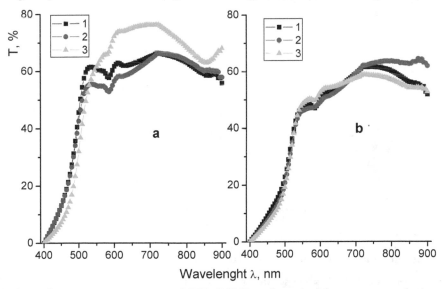

Fig. 13. Optical transmittance spectra of CSD CdS films deposited from aqueus solution: CdCl$_2$ (1); CdSO$_4$ (2), CdJ$_2$ (3) as deposited (a) and after annealing (b).

4. Solar cell performance

The n-CdS/p-CdTe HJ was fabricated and their electrical and photoelectric properties were investigated. The CdS thin films with 100 nm thickness were deposited by CSD using CdCl$_2$

cadmium chloride solution. Thin polycrystalline CdS films completely covered the substrate across the sample area, hade stoichiometric composition, was solid with a small surface macrodefects concentration ($10^7 sm^{-2}$). Typical spectral dependence of transmission of CSD CdS film is shown on Fig. 13. Resistance of is fabricated n-CdS/p-CdTe SC $R_0 \approx 10^4$-10^5 Ω at T= 300 K and was determined by the electrical properties of the p-CdTe substrates. This is coused by the resistivity of used substrates which is 2-3 orders of magnitude greater than the similar parameter for n-CdS films ($R_{CdS} \approx 10^3$ Ω). Voltage cutoff in n-CdS/p-CdTe structures, as seen in Fig. 14 is $U_0 \approx 1,4$ V and its value is close to CdTe bandgap (Landolt-Börnstein, 1999). Inverse branches of curent-voltage characteristic for anisotropic structures are well described by power dependence $I_R \sim U^m$, where the m $\approx$ 1 to U> 2, which is typical for charge carriers tunneling or inherent space charge limited currents in velocity saturation mode (Hernandez, 1998, Lamperg & Mark, 1973). Reverse current increase observed in the investigated anisotropic heterojunction with increasing voltage bias can also be caused by imperfections in their periphery.

Fig. 15 shows relative quantum efficiency of photoconversion (ratio of short circuit current to number of incident photons) $\eta(h\nu)$ spectra of CdS/CdTe heterojunction fabricated by CSD of CdS film on CdTe wafer. The $\eta(h\nu)$ spectra find out to be similar for structures fabricated on different substrates what indicate high local homogeneity of substrates and reproducibility of the CSD films properties. The sharp long wave increase of $\eta(h\nu)$ in narrow spectral range 1.4–1.5 eV for CdS/CdTe structure illumination from CdS film side is observed. Its value reach maximum in region $h\nu^m \approx 1.5$ eV what correspond to energy of direct band transitions in CdTe (Landolt-Börnstein, 1999, Aven & Prener, 1967).

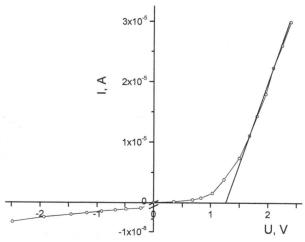

Fig. 14. Curent-voltage characteristic of n-CdS/p-CdTe HJ at 300 K

It should be notice that photosensitivity of the fabricated CdS/CdTe heterostructures maintain on high level (fig. 15, curve 1, 2) in vide region of incident photons energy. The table like part of $\eta(h\nu)$ curve confirm fabrication of the CdS/CdTe high quality heterojunction. The observed $\eta(h\nu)$ curve decrease at $h\nu \geq 2.3$ eV is similar to specular transmission spectra of CdS film used for CdS/CdTe heterostructure fabrication. The full wide on half of the maximum (FWHM) of $\eta(h\nu)$ spectra $\delta \approx 1.1$–1.2 eV in our structures is more bigger then FWHM of Ox/CdTe heterostructure (Il'chuk at al., 2000) and indicate higher quality of fabricated structures compared to known.

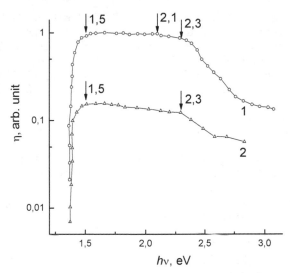

Fig. 15. Spectral distribution of the quantum efficiency of n-CdS/p-CdTe heterojunction at 300 K. Ilumination form CdS film side. The 1 and 2 curves corespond to samples with diferet thickness of CdS film.

5. Conclusions

The II-VI binary CdS compound semiconductor thin films (30–100 nm) has been successfully deposited from aqueous solutions of $CdCl_2$, $CdSO_4$, CdI_2 salts using Chemical Surface Deposition and employing the direct heating of the substrate.

The linear dependence increase of film thickness on the deposition time was experimentally demonstrated for CSD $CdCl_2$ based films. For the two other salts of the film thickness dependence on deposition time is more complex, but has a character close to linear.

Established that for growth rate <15 nm/min. chemical deposition method allows to growth solid polycrystalline CdS films with 10^6-10^7 cm^{-2} surface macrodefects concentration.

It is proved that CdI_2 based CdS film composition was close to stoichiometric, compared to films deposited from solutions of two ather salts under identical conditions.

The possibility of n-CdS/p-CdTe high quality solar cell fabrication by CSD of CdS thin film is demonstrated. High value of CdS/CdTe heterojunction photoconversion, in region limited by CdS and CdTe band gaps, in our opinion was provided by CdS deposition method.

6. References

Adams, W.G. (1877). The action of light on selenium. *Proceedings of the Royal Society*, No.25, pp. 113–117

Adirovich, E.I., Yuabov, Y.M., & Yagudaev, G.R. (1969). CdTe thin film deposition on transparent substrates. *Fiz. Tekh. Poluprovodnikov*, Vol.3, No.1, pp. 81–85

Aguilar-Hernández, J., Sastre-Hernández, J., Ximello-Quiebras, N., Mendoza-Pérez, R., Vigil-Galán, O., Contreras-Puente, G., & Cárdenas-García, M. (2006). Influence of the S/Cd ratio on the luminescent properties of chemical bath deposited CdS films. *Solar Energy Materials & Solar Cells*, Vol.90, pp. 2305–2311

Archbold, M.D., Halliday, D.P., Durose, K., Hase, T.P.A., Smyth-Boyle, D., & Govender, K. (2005). Characterization of thin film cadmium sulfide grown using a modified chemical bath deposition process. *Conference record of the thirty-first ieee photovoltaic specialists conference*, 3-7 January 2005. – USA, 2005, pp.476–479

Aspects of Heterojunction Formation. *Thin Solid Films*, Vol.387, pp.141-146

Aven, M., & Prener, J.S. (Ed(s).). (1967). *Physics and Chemistry of II-VI Compounds*, North-Holland, Amsterdam

Basol, B., & Kapur, V. (1990). Deposition of CuInSe$_2$ films by a two-stage process utilizing E-beam evaporation. IEEE Trans. Electron Dev, No.37, pp. 418–421

Basol, B.M. (1990) Thin film CdTe solar cells-a review. *Proceedings of 21st IEEE Photovoltaic Specialists Conference*, pp. 588–594, Kissimimee, USA, 21–25 May, 1990

Basol, B.M., Kapur, V.K., Halani, A., & Leidholm, C. (1991). *Annual Report, Photovoltaic Subcontract Program FY*, pp. 50

Becquerel, E. (1839). Mémoire sur les effets électriques produits sous l'influence des rayons solaires. *Comptes Rendues*, No.9, pp. 561–567

Bode, D.E., Hass, G., & Thun, R.E. (1966). *Physics of Thin Films vol. 3*, Academic Press, New York

Bonnet, D., & Rabenhorst, H. (1972) 6 % efficient CdS/CdTe solat cell. *Proceedings of 9th IEEE Photovoltaic Specialists Conference*, pp. 129–131, Silver Spring, USA

Boudreau, R.A., & Rauh, R.D. (1983). Chemical bath deposition of thin film cadmium selenide for photoelectrochemical cells. *J. Elctrochem. Soc.*, No.130, pp. 513

Calixto, M.E., & Sebastian, P.J. (1999). A comparison of the properties of chemical vapor transport deposited CdS thin films using different precursors. *Solar Energy Materials & Solar Cells*, No.59, pp. 65-74

Chaisitsak, S., Yamada, A., & Konagai, M. (2002). Preferred Orientation Control of Cu(In$_{1-x}$Ga$_x$)Se$_2$ ($x \approx 0.28$) Thin Films and Its Influence on Solar Cell Characteristics. *Jpn. J. Appl. Phys*, Vol.41, pp. 1347-4065

Chapin, D.M., Fuller C.S., & Pearson, G.L. (1954). A new silicon p-n junction photocell for converting solar radiation into electrical power. *Journal of Applied Physics*, No.25, pp. 676–678

Chen, F., & et all. (2008). Effects of supersaturation on CdS film growth from dilute solutions on glass substrate in chemical bath deposition process. *Thin Solid Films*, Vol.516, pp. 2823–2828

Chopra, K. L., & Das, S. R. (1983). *Thin Film Solar Cells*, Plenum Press, New York

Contreras, M. A., Romero, M. J., To, B., & at all. (2002). Optimization of CBD CdS process in high-efficiency Cu(In,Ga)Se$_2$-based solar cells. *Thin Solid Films, Vol.* 403–404, pp. 204-211

Estela Calixto, M., Tufiño-Velázquez, M., Contreras-Puente, G., & at all. (2008). Study of chemical bath deposited CdS bi-layers and their performance in CdS/CdTe solar cell applications. Thin Solid Films, No.516, pp. 7004–7007

Estrada-Gasca, C., Alvarez-Garcia, G., & Nair, P.K. (1993). Theoretical analysis of the thermal performance of chemically deposited solar control coatings. *J. Phys. D*, No.26, pp. 1304–1309

Estrada-Gasca, C., Alvarez-Garcia, G., Cabanillas, R.E., & Nair, P.K. (1992). Theoretical efficiency of an all-glass tubular solar collector using a chemically deposited SnS-Cu$_x$S absorber inside the inner tube. *J. Phys. D*, No.25, pp. 1142–1147

Fahrenbruch, A.L., & Bube, H. (1983). *Fundamentals of Solar Cells*, Academic Press, London

Granath, K., Bodegard, M., & Stolt L. (2000). The effect of NaF on Cu(In, Ga)Se$_2$ thin film solar cells. *Sol. Energy Mater. Sol. Cells*, Vol. 60, pp. 279-293

Green, M. (1998). *Solar Cells – Operating Principles, Technology and System Applications*, The University of South Wales

Green, M.A., Keith, E., Hishikawa, Y., & Warta, W. (2011). Solar cell efficiency tables (Version 37). *Progress in Photovoltaics: Research and Application*, No.19, pp. 84–92

Grozdanov, I. (1994). A simple and low-cost technique for electroless deposition of chalcogenide thin films. *Semicond. Sci. Technol.*, No.9. pp. 1234–1241

Hamakawa, Y. (2002). Solar PV energy conversion and the 21st century's civilization. *Solar Energy Materials and Solar Cells*, Vol.74, No.4, pp. 13-22

Hashimoto, Y., Kohara, N., Negami, T., Nishitani, N., & Wada, T. (1998). Chemical bath deposition of Cds buffer layer for GIGS solar cells. *Sol. Energy Mater. Sol. Cells*, Vol.50, pp. 71-77

Hass, G., Francombe, M.H., & Vossen, J.L. (Eds.). (1982). *Physics of Thin Films vol. 12*, Academic Press, New York

Hernandez, E. (1998). Space-charge-limited current effects in p-type $CuIn_{0.8}Ga_{0.2}Se_2/In$ Schottky diodes. *Cryst. Res. Technol*, No.33, pp. 285–289

Houser, O., & Beisalski, E. (1910). PbS film fabrication. *Chem-Ztg*, No.34, pp. 1079–1081

Huang, L., Nair, P.K., Nair, M.T.S., Zingaro, R.A., & Meyers, E.A. (1994). Interfacial diffusion of metal atoms during air annealing of chemically deposited ZnS-CuS and PbS-CuS thin films. *J. Electrochem. Soc*, No.141, pp. 2536

Il'chuk, G.A., Ivanov-Omski, V.I., Rud', V.Yu., Rud', Yu.V., Bekimbetov, R.N., & Ukrainets, N.A. (2000). Fabrication and photoelectric properties of Oxide/CdTe structures. *Semiconductor structures, interfaces, and surfaces*, Vol.34, No.9, pp. 1099-1102

Jager-Waldau, A. (2004). R&D roadmap for PV. *Thin Solid Films*, Vol.451-452, pp. 448-454

Jenny, D.A., Lofersky, J.J., & Rappaport, P. (1956). Photovoltaic Effect in GaAs p-n Junctions and Solar Energy Conversion. *Physical Review*, No.101, pp. 1208–1209

Kazmerski, L.L., White, F.R., & Morgan, G.K. (1976). Thin-film CuInSe2/CdS heterojunction solar cells. *Applied Physics Letters*, No.29, pp. 268–270

Khallaf, H., Oladeji, I. O., Chai, G., & Chow, L. (2008). Characterization of CdS thin films grown by chemical bath deposition using four different cadmium sources. *Thin Solid Films*, No.516, pp. 7306

Kitaev, G., Mokrushin, S., & Uritskaya, A. (1965). *Kolloidn. Z.*, No.27, pp. 51

Lamperg, G., & Mark, P. (1973). *Injection curents in solids*, Mir, Moscov

Landolt-Börnstein. Semiconductors; II-VI and I-VII Compounds; Semimagnetic Compounds. Landolt-Börnstein. – Berlin·Heidelberg·New York: Springer-Verlag, 1999. – Group III, Vol. 41b

Lokhande, C.D. (1991). Chemical deposition of metal chalcogenide thin films. *Mater. Chem. Phys*, No.27, pp. 1–43

Mickelsen, M.A., & Chen, W.S. (1980). High photocurrent polycrystalline thin-film CdS/CuInSe2 solar cell. *Appl. Phys. Lett.*, No.36, pp. 371

Möller, H. J. (1993). *Semiconductors for Solar Cells*, Artech House, London

Mugdur, P.H., Chang, Y.-J., Han, S-.Y., & at all. (2007). A Comparison of Chemical Bath Deposition of CdS from a Batch Reactor and a Continuous-Flow Microreactor. *J. Electrochem. Soc.*, No.154, pp. D482–D488

Nair, P.K., Garcia, V.M., Comer-Daza, 0., & Nair, M.T.S. (2001). High thin-film yield achieved at small substrate separation in chemical bath deposition of semiconductor thin films. *Semicond. Sci. Tcchnol*, Vol.10, No.16, pp. 855–863

Nair, P.K., Huang, L., Nair, M.T.S., Hu, H., & at all. (1997). Formation of p-type Cu3BiS3 absorber thin films by annealing chemically deposited Bi2S3-CuS thin films. *J. Mater. Res.*, No.12, pp. 651–656

Nair, P.K., Nair, M.T.S., Fernandez, A., & Ocampo, M. (1989). Prospects of chemically deposited metal chalcogenide thin films for solar control applications. *J. Phys. D*, No.22, pp. 829–836

Nakada, T., & Kunioka, A. (1999). Direct evidence of Cd diffusion into Cu.In,Ga.Se2 thin films during chemical-bath deposition process of CdS films. *Appl. Phys. Lett*, Vol.**74**, pp. 2444-2446

Oladeji, I.O., & Chow, L. (1997). Optimization of Chemical Bath Deposited Cadmium Sulfide Thin Films. *J. Electrochem. Soc.*, Vol.144, No.7, pp. 2342-2346

Ortega-Borges, R., & Lincot, D. (1993). Mechanism of chemical bath deposition of cadmium sulfide thin films in the ammonia-thiourea system. *J. Electrochem. Soc.*, No.140, pp. 3464-3473

Palatnik, L.S., & Sorokin, V.K. (1978). *Materialovedenie v mikroelektronike*, Energia, Moscow

Qiu, S. N., Lam, W. W., Qiu, C. X., & Shih, I. (1997). ZnO/CdS/CuInSe2 photovoltaic cells fabricated using chemical bath deposited CdS buffer layer. *Appl. Surf. Sci.*, Vol.113/114, pp. 764-767

Rau, U., & Scmidt, M. (2001). Electronic Properties of ZnO/CdS/Cu(In, Ga)Se2 Solar Cells -

Reddy, T.A., Gordon, J.M., &. de Silva, I.P.D. (1987). Mira: A one-repetitive day method for predicting the long-term performance of solar energy systems. *Sol. Energy*, No.39, pp. 123–133

Reynolds, D.C., Leies, G., Antes, L.L., & Marburger, R.E. (1954). Photovoltaic effect in cadmium sulfide. *Physical Review*, No.96, pp. 533–534

Rincon, M.E., Sanchez, M., Olea, A., Ayala, I., & Nair, P.K. (1998). Photoelectrochemical behavior of thin CdS, coupled CdS/CdSe semiconductor thin films. *Solar Energy Mater. & Solar Cells*, No.52, pp. 399–411

Romeo, N., Bosio, A., & Canevari, V. (2003). The role of CdS preparation method in the performance of CdTe/CdS thin film solar cell. *3rd World Conference on Photovoltaic Energy Conversion*, 11–18 May 2003, Osaka, Japan, pp. 469–470

Romeo, N., Bosio, A., Tedeschi, R., & Canevari, V. (2000). Back contacts to CSS CdS/CdTe solar cells and stability of performances. *Thin Solid Films*, No.361-362, pp. 327–329

Rothwarf, A. (1982). *Proc. 16th IEEE Photovoltaic Specialists Conf.*, San Diego, 1982, CA, IEEE, New York, pp. 791

Savadogo, O., & Mandal, K.C. (1993). Low-cost technique for preparing n-Sb2S3/p-Si heterojunction solar cells. *Appl. Phys. Lett.*, No.63, pp. 228

Savadogo, O., & Mandal, K.C. (1994). Low cost schottky barrier solar cells fabricated on CdSe and Sb2S3 films chemically deposited with silicotungstic acid. *J. Electrochem. Soc.*, No.141, pp. 2871

Soubane, D., Ihlal, A., & Nouet, G., (2007). *M. J. Condensed Matter*, Vol.9, pp. 32-35

Stevenson, R. (2008). First Solar: Quest for $1 Watt. *Spectrum*, Vol.45, No.8, pp. 22–33

Svechnikov, S.V., & Kaganovich, E.B. (1980). CdSxSe1-x photosensitive films: Preparation, properties and use for photodetectors in optoelectronics. *Thin Solid Films*, No. 66, pp. 41–54

Sze, S. M. (1981). *Physics of semiconductor devices (2nd edn.)*, John Wiley & Sons Inc., New York

Tiwari, S., & Tiwari, S. (2006). Development of CdS based stable thin film photo electrochemical solar cells. *Solar Energy Materials & Solar Cells*, Vol.90, pp. 1621–1628

Tuttle T.R., Contreras M.A., et al. (1995). *Proc. SPIE*, Vol.2531, SPIE, Bellingham, pp. 194

Vossen J.L., & Kern W. (1978). *Thin Film Processes*, Academic Press, New York

Wagner, S. (1975). Epitaxy in solar cells. *Journal of Crystal Growth*, No.31, pp. 113–121

Wolf, R. (Ed). (1975). *Cadmium Sulphede Solar Cells*. Applied Solid State Sciences, vol. 5, Nev York

10

What is Happening with Regards to Thin-Film Photovoltaics?

Bolko von Roedern
National Renewable Energy Laboratory
United States of America

1. Introduction

The advances and promises of thin-film photovoltaics (PV) are much discussed these days, typically using the viewpoint that a picked technology and process approach would provide "the" solution to many problems experienced implementing PV commercialization. In 2009, a thin-film PV company, First Solar, garnered world-leadership as a PV company, being the first company to produce or ship more than 1 GW of PV modules in a single year. This makes it timely to discuss the advantages and limitations of thin-film PV technology, as compared to the currently prevailing crystalline Si PV industry. Traditionally, the following technologies are considered constituting "thin-film PV:"

1. CdTe PV
2. CIGS PV (or copper-indium-gallium diselenide)
3. a-Si:H (and nc-Si:H nanocrystalline or "micromorph" silicon films)
4. less than 50 micron thick crystalline Si films

In the amorphous silicon (a-Si:H) based category, several approaches are pursued, ranging from amorphous silicon single junction modules to spectrum splitting multijunction cell structures using either a-SiGe:H cell absorbers or a-Si:H/nc-Si:H multijunctions. Pros and cons will be given for these different approaches that lead to this multitude of device structures. It is argued that as long as the advantages of the aforementioned materials are not understood, it would be difficult to "design" materials for more efficient solar cell operation.

This review will recap what is currently known about these materials and solar cell devices, keeping in mind that there will always be some unexpected "surprises," while there were many other approaches that did not result in anticipated cell/module performance improvements. This knowledge leads the author to ask the following question: "Was improper implementation or inadequate process choice responsible for the lack of solar cell/module performance improvement, or was the expectation for improved device performance or decreased device cost simply not warranted?"

The chapter of this book is written such as to not prejudge an outcome, i.e., an a priori assumption that a given measure would result in a commensurate expected performance improvement. The impact (i) of an improvement is broken down into probability (p) of achieving a projected improvement times the effectiveness (e) of such improvement, where

$$i = p \cdot e$$

It is of interest to note that while impact is costed and/or priced by many companies, the right hand side of the above equation also has associated cost elements associated with effectiveness e plus an estimated probability p. Probabilities (p between 0 and 1 or 0% and 100%) are often assumed to be either 0% (for an unsuccessful project) or 100% (for a successful project), with the benefit of hindsight. This is true only with the benefit of hindsight, forward looking probabilities should be estimated and accounted for as accurately as possible. In financial terms, a probability between 0 and 1 should be accounted for by applying appropriate financial discounts to probabilities falling outside the extreme values, 0 or 1. Instead, often p=1 is being "assumed," but strictly speaking, this is inadmissible in forward-looking situations. Whenever p is increased at the expense of e, the total benefit for i may not be achieved as planned. Typically, p has to be empirically assessed, which is important for appropriate financial "discounting" leaving much room for discussion as to what value (between 0% and 100%) to assign to p. The foregoing statement is valid for all PV technologies (not just thin-film PV), but in the following, mainly elucidated picking thin-film PV examples. This chapter does not want to chime in on a debate about what appropriate probabilities or discount factors should be used, but rather serve as a reminder to the fact that projected probabilities occur with less than 100% probability.

2. Status and challenges for CdTe based solar cells and modules

In the year 2009, a company relying on producing CdTe based PV modules, First Solar Inc., became the World's largest photovoltaic (PV) company, producing about 1,100 MW of PV modules. Its production costs per Watt were quite low by industry standards. In 2010, direct manufacturing costs of less than $0.8/W were reported by First Solar. First Solar modules are 120 cm x 60 cm in size and were reported in 2010 to generate between 70 and 82.5 Watts under standard testing conditions, resulting in commercial module efficiency levels on the order of 10% to 11.5%. Time will tell how much room there is to further enhance power ratings and commercial module efficiency. It can be expected that in the foreseeable future, First Solar will remain among the top World Producers of PV modules. The CdTe device is a true thin-film device consisting of a TCO-coated (typically, SnO_2) glass superstrate, a CdS junction partner layer, an active CdTe layer, an often proprietary back contact, packaged in a hermetically sealed package. First Solar buys SnO_2-coated superstrates, uses vapor transport deposition (VTD) for the CdS and CdTe layers, and applies a proprietary back contact and cell series interconnect to the device structure.

Champion CdTe cells have achieved in excess of 16% efficiency (Green et al, 2011). It is of concern to some researchers that this champion cell was reported already some 10 years ago and has not improved since. The compound semiconductor CdTe has a tendency to grow and sublime stoichiometrically when exposed to high temperature. Instead of using vapor transport deposition used by First Solar, many R&D efforts use "Close-Space Sublimation" (or CSS) to deposit the CdTe layer. It appears that the deposition method for the CdS junction partner layer is not of as great importance as in the case of CIGS solar cells, where frequently chemical wet deposition schemes are used for depositing the CdS layer which is only about 100 nm thick, because that deposition method produces the greatest and most reproducible performance. CdTe layers deposited at the highest temperature compatible with the soda-lime glass superstrates typically result in the greatest device efficiency. However, other CdTe deposition schemes, most notably electro-deposition, also resulted in PV modules exhibiting substantial efficiency and performance (Cunningham et al, 2002). It was, however, found that a critical $CdCl_2$-anneal step is crucial to achieve best solar cell or module performance (McCandless, 2001). Anneal temperatures on the order of 400 °C are typically used after the $CdCl_2$ exposure. For industrial production rates, it is important to limit the time for such anneal step in order to achieve an appropriate throughput. Looking at current commercial throughput rates, one has to conclude that this is possible. It was also attempted to substitute this $CdCl_2$ anneal step (where $CdCl_2$ is often applied as an aqueous solution] with a gaseous anneal step using HCl dry gas (McCandless, 2001). While this approach resulted in similar results as the aqueous $CdCl_2$ anneal step, a superiority using this "dry" process could not be established .

CdTe cells can be made stable and lasting, but not all production schemes result in stable cells. It was reported that excessive reliance on the $CdCl_2$-anneal step to obtain the highest cell or module efficiencies often led to less stable devices (Enzenroth et al., 2005), with processes leading to the highest pre-anneal efficiency often resulted in the most stable manufacturing recipes. It is now known that Cu, applied to many back-contacting schemes, is correlated with the stability of CdTe cells. While it has been established that "too much" Cu results in unstable cells, some rather stable cell deposition schemes were developed that use Cu-doped back contact recipes. The degradation process shows a mixture of diffusive and electromigration behavior (Townsend et al., 2001). Alternatives to using Cu for the back contact were developed (e.g., P-doping, N-doping) (Dobson et al. 2000). These 'Cu-free' recipes also showed instabilities and did so far not improve cell performance over that achieved with stable Cu-containing back-contact recipes. Perhaps, it is a flaw to ask: "Is Cu in the back contact good or bad for cell stability?" The appropriate question may well be: "When is Cu good, when is it bad, and when is it irrelevant for cell performance and stability?"

While all commercial CdTe solar modules are currently fabricated in a superstrate configuration (using a glass superstrate), the question has been posed whether such process could be inverted and/or be applied to flexible substrates. Flexible substrates (like polyimide foil) limit the temperature that can be applied during the position process. Also, the issue of low-cost hermetic packaging of such transparent foils has to be addressed in greater detail in a cost-effective manner. Because glass-encapsulated PV works, the cost of glass (on the order of $10/m^2$ for a single sheet) can often be used as a cost-guideline for terrestrial flexible packaging schemes for power modules. It is clear at this juncture that CdTe PV and CIGS PV have greater moisture sensitivity than many Si PV schemes, requiring a more hermetic seal than Si PV might require. A point of research continues to be the "edge delete" for modules. Typically, SnO_2 coated superstrates are coated with all layers of the entire glass surface. A fast removal of such films, including the SnO_2-layer, along the module edges is required. For CdTe modules, often rather crude methods (like bead-blasting or using grinding wheels) are employed for this "edge delete" step were employed. The drawback of employing these methods is that glass surfaces are damaged using such processes, resulting in greater water penetration rates from the module edges. Also, such processes tend to weaken the glass. However, less damaging edge delete techniques like laser ablation methods are rapidly becoming feasible and more cost effective.

In order to make a monolithically interconnected module, cell "strips" have to be created that carry CdTe currents through the SnO_2. Typically, 1 cm-wide cell strips are used for CdTe modules. These strips require 3 scribes sometimes labeled P1 (SnO_2), P2 (semiconductor layer), and P3 (back-contact) scribe line. The area including and between scribes P1 and P3 is electrically "dead" and does not contribute to module power, hence reducing the total area module efficiency. Therefore, scribe lines should be narrow and close to each other, which requires a good parallel alignment of the scribe lines with each other. With the advancement of laser technology, all of these scribes are often achieved by laser scribing. CdTe (and CIGS) cells can also be scribed with a mechanical stylus, and sometimes, lift-off techniques were used for the P3 scribe by printing a lift-off paste to segment the cell's back contact. CdTe modules can be scribed in a picture frame or landscape format. First Solar scribes in a picture frame format, arguing that high module voltages would reduce resistive (I^2R) losses in the dc module wiring. However, it was also discovered that modules are installed with a maximum string voltage of 600V (dc, in North America, 1000V in Europe), leading to relatively short strings for high-voltage modules. Realizing this, for its series 3 modules, First Solar has reduced the voltage, resulting in lower voltage (and greater current) PV modules. Other CdTe companies have elected to scribe in a landscape format.

Research activities for CdTe cells and processes concern themselves with achieving a greater open-circuit voltage (V_{OC}), greater stability, and more repeatable solar cell processing. While the CdTe semiconductor possess nearly the ideal band gap for absorbing the solar spectrum in a single junction device (about 1.5 eV), V_{OC} is limited to approximately less than 900 mV for champion cells,(about 750 - 800 mV per cell for commercial devices), well below values that were achieved for high efficiency GaAs solar cells (V_{OC} of about 1200mV in "champion" cells) where the semiconductor absorber has a very similar band gap near 1.5eV. Investigation of back contacts and device stability is sometimes hampered by the proprietary nature used by industry for these processes. Also, the role of impurities (oxygen, water vapor) and the process when and how these impurities are added are currently poorly understood.

Long-term concerns for CdTe PV are a perceived toxicity (Cd-containing compounds) and the availability of Te. While Te availability is not a problem now, it may become so after multiple terra-Watts of CdTe PV have been produced. A known mitigation scheme for incorporating less Te (and Cd) into a cell would be to make the absorber layer thinner. Unfortunately, as the absorber thickness is reduced to values below 1.5 microns, an often precipitous decrease in cell fill factor and V_{OC} were observed. For some solar cell processes, a more gradual decrease of these cell parameters is observed even as thicker absorber layers are thinned. Because absorber material costs are not a significant manufacturing cost factor, manufacturers are reluctant to sacrifice performance by making thinner absorbers, hence the development of thin absorber cells is currently only infrequently pursued. There comes a point when very thin absorber cells would also loose current density due to incomplete light absorption, but in a direct band gap thin-film semiconductor this would only happen for absorber thicknesses below 1 micrometer. Further, as the a-Si:H and nc-Si:H PV communities have shown, it may be possible to mitigate such current loss by employing optical enhancement techniques (Platz et al. 1997).

3. Status and challenges for CIGS based devices

Champion CIGS Cells have been reported near 20% cell efficiency (Green et al, 2011). It is remarkable that (a) 2 different groups on two continents (National Renewable Energy Laboratory, NREL and Center for Solar Hydrogen, ZSW) have achieved this efficiency level, and that (b) different material compositions all can achieve high efficiency cells (Noufi 2010). The record cells were mostly made by a process call co-evaporation. Typically, this process has multiple "stages" involved, finishing devices with a Cu-poor (or In-rich) surface layer (Gabor et al. 1994). This process has also been adopted for CIGS module manufacturing. Other processes used for commercial module fabrication are sputtering and (time-consuming!) selenization using H_2Se gas, various hybrid processes, electro-deposition, and nano-particle precursor inks. Only time will tell if the latter deposition processes can achieve the same performance as the co-evaporation process can? There are currently different schools of thought as to why best solar cell results are obtained using these multi-stage processes. Some people argue that the Cu-poor surface phase is a perfect ordered vacancy compound (Schmid et al. 1993), while other researchers believe that a non-perfect Cu-deficient surface layer can enhance CIGS solar cell performance (It may be instructive to compare this issue to the crystalline Si PV case. Traditionally, this PV technology has used monocrystalline and multicrystalline Si wafers. While several promoters have some understanding that there is an efficiency difference between mono-Si and multi-Si based technologies, some Si advocates say that all Si cells "should" have the same efficiency potential.)

Nano-particle approaches have been promoted based of the promise that the absorber properties could be fixed in the ink precursor. Nevertheless, the scale-up of nano-particle precursor deposition approaches has also shown significant variation in output power. Researchers typically have the uniformity of a semiconductor absorber layer in mind when looking at enhanced control scheme, thereby neglecting the "junction-uniformity" upon scale up, which can be observed in any commercial manufacturing process even when the absorber properties remain constant upon deposition area scale-up and/or throughput.

This author ranks the probability as quite low that Se could be added in a "fast" process to metallic precursor layers. Past work was carried out along these lines (Attar et al. 1994) . Similarly, advantages of CuSe or InSe precursors have not as yet been demonstrated to lead to high solar cell efficiencies (Anderson et al. 2003). In addition to films made by the former process having problematic mechanical film properties (flaking), rapid post-deposition selenization approaches have also not yet lead to great solar cell efficiency. This observation currently necessitates handling a high vapor pressure Se (relative low temperature) Se evaporation source and low vapor pressure Cu evaporation source (relative high evaporation temperature) in the same vacuum system.

CIGS PV showed the last significant "win-win" situation in PV when it was suggested (for reasons of lowering manufacturing cost) to change substrate material from using boro-silicate glasses to soda lime (ordinary window) glass. What was not anticipated was that such switch also increased the cell performance obtained. It is now understood that controlled addition of Na can enhance the performance seen in CIGS cells. In fact, Na addition was essential for making high-efficiency CIGS cells on metal foils a reality. The reasons for this advantage are poorly understood, but the observation is overwhelming that Na can improve CIGS solar cell performance.

The CIGS cell typically consists of the following structure: Glass/Mo-film/multi-stage-CIGS/CdS/TCO. Since a finished cell can be exposed only to moderate temperature (<200 °C, perhaps <150 °C), sputtered ITO or ZnO or LPCVD (Low Pressure Chemical Vapor deposited) ZnO are typically used as the TCO. The Mo-film and the TCO deposition processes may use more than one deposition process for fabricating such layer (e.g. sputtering condition). When using co-evaporation for the CIGS deposition process, the best performance results are obtained when substrate temperatures during the deposition process are high, approaching the softening point of glass. The CdS layer, for high performing CIGS cells and modules, uses a wet (CBD chemical bath deposition) process for a thin (100 nm thick) CdS layer. For modules, scribing the p(1) through p(3) scribe lines can involve laser and/or mechanical methods (Tarrant & Gay, 1995). Because of a higher current density in CIGS (typically, 33 mA/cm^2 ± 15%) cell strips are typically only 5 to 6 mm wide. For such cells, scribing tolerances are particularly important for minimizing the non-contributing module area.

Many commercial CIGS modules are currently fabricated on rigid glass substrate/cover glass structures, limiting moisture ingress to the module perimeter. Even these structures initially had problems passing the damp heat (1000 hours at 85% relative humidity, 85 C) tests. This suggests that CIGS cells are more moisture sensitive than modules made using Si solar cells. Some commercial CIGS manufacturers fabricated on flexible metal foil material have therefore designed their cells as Si cell replacement to be packaged within glass sheets. The question has been posed whether such process could be inverted and/or be applied to flexible substrates. Flexible substrates (like polyimide foil) limit the temperature that can be used to deposit the CIGS films, but allow monolithic (scribed) integration of the module, while stainless steel substrates allow the use of higher deposition temperatures, but, because they are conductors, not the monolithic interconnection. Typically, PV made on metal foils is "slabbed" into individual solar cells, giving up some advantages of a roll-to-roll fabrication process.

In order to increase the humidity tolerance, it is presently not clear whether to make the solar cell more tolerant to moisture or whether to lower the water transmission rate of the module package. It is known that the ZnO layer used as the top contact by some entities deteriorates upon moisture contact. Some groups therefore work on replacing the TCO material. On the other hand, it is also known that there can be degradation for CIGS cell recipes that use an ITO instead of a ZnO contact for CIGS cells, and that other technologies (like a-Si or a-Si/nc-Si technologies) have achieved acceptable stability using ZnO for top and/or bottom solar cell contacts. It is somewhat likely that there is not a single cause or mechanism for moisture sensitivity, and that CIGS PV will be more sensitive to moisture than Si–based PV. This leaves the question how cost-competitive flexible CIGS is for power generation. Such competitiveness will require a light-weight, flexible and optically transparent low-cost moisture barrier. Acceptable barriers may exist as commercial prototypes, but commercial cost for such foils is not clear. If these foils were significantly more expensive than glass, the advantage of flexible CIGS PV could be diminished.

The long-term stability of CIGS is acceptable, depending on details of device processing and the quality of the package. Having been discovered some time ago, "transients" in CIGS-based devices are poorly understood. If finished solar cells or modules are exposed to moderate heat in the dark (<150 °C, for example when modules are laminated), a power loss is often (but not always) observed. Such behavior is currently not predictable. Often, but not always, the power loss recovers when the module is exposed to natural or artificial light. These "transients" may change as modules age and pose a problem for qualification tests, specifying a pre-and post stress power variations that could be larger than stress induced power losses. For some CIGS pilot production modules, it was found that such transient loss effects were on the same order as stress or deployment induced losses. The question is to what degree recovery can be relied upon to achieve performance predictions that on average are correct?

Some tests (like the 85/85 test) heat the modules in the dark. Because of this behavior, the qualification test for modules utilizes the manufacturer's labeled module power rather than the measured module power as the criterion for power loss upon stressing. CIGS (and all) thin-film modules are tested using the IEC 61646 accelerated testing specifications. One manufacturer exposed CIGS modules with questionable lamination power losses to actual sunlight to ascertain the amount of recovery.

Long-term potential limitations to CIGS PV are the limited availability of In metal. The use of In could be reduced by manufacturing thinner cells than the thicknesses used today. However, experimental and commercial reality is similar to what has been said about thin CdTe solar cells above, because materials cost for the semiconductor layer currently are low, typically best performance, not minimum thickness is used for commercial activities. It is also unclear if a competing technology, flat panel displays, will continue to use In (ITO) or will switch to a different TCO material. Being limited by In availability is not expected to be a problem until terawatts of CIGS PV modules have been fabricated. Another potential problem is customer acceptance. CIGS cells use a small amount of CdS in the buffer. Several entities have therefore developed alternative buffers to CdS (Contreras et al. 2003). Such work may be successful (but no performance improvements were yet found because of using alternative buffers), and it is of interest to note that a similar wet deposition process for best alternate junction partners also uses CBD. There are also efforts to develop CIGS-solar cells using earth-abundant non-toxic materials only. This requires replacing the In (and perhaps Ga) used in CIGS solar cells. A popular candidate is currently Zn ("CZTS" cells), and efficiencies near 9% were reported for such cells (Todorov et al. 2010). Using such alternative materials suffers from the fact that the "secret" of CIGS solar cell operation is not understood (why the device optimizer has to do what he has to do in order to attain high efficiency solar cells, why In, Ga and CBD CdS work extremely well). Researchers focus on materials that have appropriate optical properties, but appear to miss out on the important relevant electronic differences between CIGS and alternative materials.

Research issues for CIGS based solar cells are: Understanding the difficulties scaling up current champion cell recipes to commercial size, understanding the befits of incorporating Na into cell, understanding the stoichiometric requirements (In to Ga to Cu to Se concentration ratios, in combination with other parameters such as solar cell thickness, chemistry of buffer layers etc.), understanding 'transients' in solar cells, understanding the 'secrets' of In, Ga, Cu, Se, and Na required for achieving champion-level efficiencies, developing alternative buffer layers, and understanding how V_{OC}, FF and J_{SC} losses could be mitigated in cell using absorbers <1 micrometer thick. There is less focus on the quality of the back contact, but unless Mo is used as the contacting layer, cell results are typically much poorer. The secret of the Mo use should be part of understanding why current champion cell recipes have to be made the way that they are being made.

4. Status and challenges for amorphous silicon and micromorph solar cells and modules

Amorphous silicon constituted the first commercial thin-film PV module product. The process of making amorphous silicon solar cells and modules was first invented by the RCA and Energy Conversion Devices (ECD) laboratories (Catalano et al. 1982, Izu et al, 1993). There was also a strong push by Japanese Companies (Sanyo, Fuji, Cannon, Sharp, to name a few) for commercializing this PV technology. At the time, both power and consumer products were being developed.

Spectrum splitting multijunction solar cells were invented in Japan (Kuwano et al. 1982) and consequently developed at ECD (later, doing business under their Uni-Solar brand name) and Solarex (later doing business as BP Solar, but in 2002, pulling out of all thin-film PV activities), and also in Japan and Europe. For a while, it was believed that this was the easiest pathway to achieving high-efficiency low-cost solar cells and modules. While multijunctions offer a theoretical efficiency advantage, the practical advantages are of a lesser degree. This is because in case of the a-Si-based multijunction cells, the subcells of the stack are not perfect in terms of their I(V) parameters. This has a beneficial aspect for energy generation, because as long as non-ideal subcells are stacked, one cannot invoke ideal mismatch factors when calculating mismatch for the stack (Chambouleyron and Alvarez, 1985). In fact, for multijunction III-V-based solar cells, it was shown that by managing the current flow through the stack (limiting the current by the top-cell), fill factors of the stack can well exceed the fill factor of the weakest cell in the stack (Wanlass & Albin, 2004).

Fabrication of a-Si:H solar cells and modules uses plasma enhanced chemical vapor deposition (PECVD). Typically, silane gas (SiH_4) (germane gas GeH_4 for a-SiGe:H layers) is piped into a deposition chamber near 1/1000 of one atmosphere, and by applying an rf frequency, hydrogenated amorphous silicon (hydrogenated amorphous silicon germanium alloy) layers are deposited. In many instances, the frequency of 13.56 MHz set aside for such applications was used to excite such plasma, but in the 1980s, it was reported that using higher frequencies could produce a-Si:H films and solar cells with slightly improved properties and/or higher deposition rates (Shah et al. 1988). The higher frequency deposition has been adopted by a few commercial companies. Amorphous silicon can be doped, typically with phosphorus or boron, by adding a phosphorus or boron containing gas to the gas mixture. Typically phosphine (PH_3) is used for n-type doping, while for p-type doping, B_2H_6, BF_3, and $B(CH_3)_3$ have been investigated among other doping gases.

The a-Si:H cells and modules are available in both substrate and superstrate configurations. Due to the relative low deposition temperature (200 °C or less) the choice of substrate material is less driven by temperature capabilities, but rather by issues like substrate availability, cost, and commercial handling issues. Commercial cells are illuminated through the p-doped contact and are hence termed n-i-p structures in substrate configuration or p-i-n structures for superstrate configurations. For superstrate configurations, a commercial or in-house prepared TCO layer is coated with one or more p-i-n sequences. Illumination through the p-type contact is clearly enhancing cell performance. Glass superstrate modules are typically scribed and interconnected into 1 cm-wide cell strips, commonly using laser scribing and welding methods. When conductive substrates (like Uni-Solar's stainless steel) are used, individual cells are cut from the substrate. Methods have been found to contact the top-contact TCO layer in such cells to extract the substantial currents from large-area cells.

Amorphous silicon (a-Si:H) PV went to its so far highest market share in 1988, thereafter losing market share because a resurgent activity in crystalline Si PV and because a-Si:H based module efficiencies were quite low and did not achieve stabilized efficiency levels that were predicted then (15% efficient module efficiency was predicted to be achievable by the late 1990s). Amorphous silicon suffers from so called Staebler-Wronski degradation. The a-Si:H based solar cells and modules are made with greater "initial" efficiency at modest deposition temperature (say 200 °C), but when the devices are exposed to light, a reduction of power (and all parameters like V_{OC}, FF, and J_{SC}) typically occurs. The exact amount of such loss depends on the details of device fabrication, but the effect is significant (say typically for commercial devices on the order of 30%). The Staebler-Wronski degradation can mostly be removed by annealing (for one hour or so) at temperatures of 130 C, but this temperature is greater than the normal operating temperature of PV modules and could damage module components. The strongest tool for mitigating such degradation is keeping the intrinsic-a-Si:H absorber layer thin. In 1990/1991, the US a-Si program therefore asked that only "stabilized values" for material properties and solar cell efficiencies should be reported. The stabilization procedure was specified as light-soaking under one-sun light intensity for 1000 hours at a sample temperature of 50° C (Luft et al. 1992). This change had two consequences: (1) a reduction of cell efficiency values as initial value were previously reported; (2) establishment and study of light-soaking in the major amorphous silicon laboratories. While this procedure is now followed by most commercial manufacturers, it is tempting to report better initial values, which is sometimes done. There was a debate whether or not Staebler-Wronski degradation could be entirely eliminated. To date, no elimination scheme has proven successful, but, as mentioned earlier, the magnitude of the effect can be controlled. While for many years it was believed that initial and stabilized performance scaled, it is now clear that smaller initial performance may result in greater stabilized performance and vice versa.

Large-area PECVD deposition may result in non-uniform deposition because different amounts of electric field are available on the rf cathodes of the typically capacitive coupled flat plate reactors used, because the wave length of the rf frequency and the physical dimension of the electrodes become of comparable magnitude. This problem typically becomes larger when higher frequencies (smaller wavelengths) are used to excite the rf (or vhf) plasma. Another source of non-uniformity arises from the fact that the feed-in distribution ratios of the precursor gases (SiH_4, GeH_4 and H_2) can change (because of consumption) in large area systems. It should be noted that on the other hand, such consumption can lead to desirable grading, say of the Ge-content in a-SiGe:H layers (Guha et al. 1988). These issues can all be overcome, by using the appropriate or segmented electrodes and gas feed-ins for large-area deposition. In the early 1990s, SERI (Solar Energy Research Institute, before the organization became NREL in 1991) specified attaining a certain amount of thickness uniformity (typically, ±5%) for large-area a-Si:H deposits in its subcontracts. These uniformity specifications were typically met, but what wasn't realized then is that the conditions used for meeting the uniformity criteria may not have been the same leading to the most efficient modules.

The following points may be important to assess degradation mechanisms: (1) there are interrelated "slow" and "fast" components to the solar cell degradation (Lee et al. 1996); (2) wrong fundamental degradation models could be the culprit for not being able to eliminate or minimize degradation, resulting in inadequate stabilization and "unexpected" degradation of commercial module product. High light intensity and low exposure temperature as well as process details like hydrogen dilution can favor the formation of fast (or 'easy to anneal') degradation (Lee et al. 1996, von Roedern & DelCueto, 2000)). Operating temperatures of an a-Si:H module could affect the annealing and stabilization process. Hence, a typical a-Si:H arrays show greater efficiency in the summer than in winter. This behavior is opposite to many other PV technologies, where efficiencies during summer are lower than during winter because higher operating temperatures (such temperature behavior also holds for a-Si:H modules) results in lower module voltages for the same radiation level, hence lower module power. For a-Si:H modules, the annealing effect (increasing efficiency) must be often more significant than the temperature effect (lowering efficiency). It is of importance to note that degradation and continued outdoor exposure affects the temperature coefficients observed. Typically for a-Si:H modules, T-coefficients become less negative, sometimes even positive after prolonged exposure. A detailed study how degradation affects the amount of degradation that is observed has been published (Whitaker et al. 1991).

There was a resurgence of a-Si:H activities after the year 2005 when big companies entered the a-Si module arena by making or adapting deposition lines for a-Si:H-based PV modules. Many researchers believed that this could result in a renaissance for a-Si:H PV. However, the question should be answered: how could this be the case when these companies used the same PECVD process that was researched for over 30 years for the deposition of a-Si:H PV modules? Several a-Si:H PV companies have recently given up on this technology. Is this because the economical circumstances were not right, or is it because performance and cost expectations for such modules could not be met? The reader must draw his or her own conclusion on this.

In the 1980s, it was proposed that changing the radio-frequency (to values greater than 13.56 MHz) could change the properties of a-Si:H and also facilitate the growth of nanocrystalline thin film (nc-Si:H) layers. The nc-Si:H layers can be grown when there is a high hydrogen dilution of the gas fed into a PECVD system (typically>98% H_2). Subsequently, nc-Si:H layers were investigated as absorber layers for a-Si:H-based solar cells. For more than 10 years, it is known that layers resulting in the highest solar cell efficiency are "mixed phase" (nanocrystallites of relative small size, typically << 50 nm in size), rather than those involving the largest grains and almost no "amorphous tissue" (Luysberg et al., 2001). Like for a-SiGe:H, it was found that the properties of nc-Si were not "quite good" enough for use in single junction solar cells. Hence, these layers are typically used as a-SiGe:H replacement in spectrum slitting multijunction solar cells. One group termed the word "micromorph" for such solar cells.

The micromorph solar cell constitutes a conundrum for the solar cell optimizer. Multijunction solar cells are to be approximately 'current matched.' A common target value for the current density in champion tandem multijunction solar cells is 13 mA/cm^2. This value is difficult to attain in the stabilized thin a-Si:H top junction. It remains to be seen if the field, if sustained, will gravitate to optically enhanced solar cell (Platz et al., 1997) or if a triple junction solar cell structure (having a current density of 8.7 mA/cm^2) will prevail. Since nc-Si:H absorber layers were developed later than a-SiGe:H, some promoters projected a greater efficiency potential for such layers than for a-SiGe:H absorbers. In reality, cell performance for multijunction cells containing a-SiGe:H and nc-Si:H is about the same (approximately 12% stabilized "total-area" efficiency). It has been suggested that the crystalline nature of the nc-Si:H layer would result in deposition rate independent properties of the nc-Si:H solar cell. Unfortunately, experimental observations could not support such prediction, nc-Si deposited at higher deposition rates (say >2 nm/sec) shows a significant loss in solar cell efficiency. Since optimum nc-Si:H cells are 1.5 to 2 microns thick (compared to 0.2 micrometer thick a-SiGe:H absorbers), deposition times for nc-Si:H absorbers are typically longer. This poses another decision for the solar module optimizer: Should one "trade" consumable cost (GeH$_4$ gas is expensive!) for even higher equipment capital cost ? The statement that currently a-SiGe:H and nc-Si:H solar cells and modules would have about the same efficiency is sometimes controversial (Yan et al., 2007). (An article in Photon International reported that while "micromorph" modules had a higher efficiency than a-Si:H modules, the plant size for micromorh deposition equipment was also greater than for a-Si:H module deposition. If the same amount of equipment was used, micromorph production will result in a lesser annual output than producing pure a-Si:H modules with the same deposition equipment.)

Some PV technologies have come under attack for using poisonous or harmful gases. There have also been reports that the use and release of system etching gases (NF_3 is typically used) could make Si-based PV less environmentally friendly, since etching gases like NF_3 possess a green-house gas potential about 20,000 times greater than CO_2. Photon International estimated that for an a-Si:H module factory, the greenhouse gas "pay-back" time could be twice as long as the energy pay-back time (pay-back time characterizes the avoided greenhouse gases or energy that is used to produce a PV system including the modules). The energy pay-back time for an a-Si:H PV array is on the order of 1 year. For crystalline Si PV the same issue may arise, as some PECVD systems used to deposit a "fire-through" a-SiN$_x$:H antireflection layers also use PECVD for depositing the a-SiN$_x$:H films in etch-cleaned PECVD chambers using NF_3. There are ways to mitigate the emission of NF_3; (1) avoid the use of NF_3 cleaning, or (2) use alternatives for etch-cleaning chambers like on-site generated F_2.

5. Status and challenges for crystalline silicon film solar cells and modules

It is intriguing to use crystalline Si films to make Si PV. There is a problem that when depositing such films, silicon may become loaded with impurities from the substrate material used. At one point in time, it was thought that this could be overcome by using Si as a substrate, for example, a Si ribbon material grown quickly. This approach has not proven successful, presumably because the crystalline Si films grow with different rates epitaxially on different substrate crystalline orientations. There is also the quest for using lower temperature substrate materials, but even a solid state recrystallization requires temperatures that exceeds soda-lime glass softening temperatures. Many groups using crystalline Si films have used, with some success, heavily doped mono-crystalline wafers, mullite (alumina) derivatives, pure graphite, multi-crystalline Si films, or specialty glasses to achieve deposition or recrystallization of crystalline Si films. For some of the foregoing substrate choices, differences in the thermal expansion coefficients of silicon and the substrate material can result in additional issues that need to be resolved. The Solar program of the US Department of Energy (DOE) projected that in the 1980s, Si PV would transition from wafers to films. Such transition, however, has not yet happened, because films still result in a rather low solar cell efficiency compared to wafer Si. Most people define film silicon less than 50 microns thick Si film on a foreign (non wafer Si) substrate as thin-film PV.

One device issue is the small voltage that is achievable using thin Si films. Values for V_{OC} near 600 mV have been reported, but typical values are lower than that, perhaps on the order of 500 mV or less. The low voltages and fill factors are a universal observation for thin cells, but many researchers focus on short-circuit current density (J_{SC}) for thin–absorber cells. This leads to the following question: Should one first tackle a loss of V_{OC} and FF in thin absorber cells, or should one begin tackling short-circuit current densities? In 1998, Dr. Jürgen Werner summarized Si film solar cell observations by plotting grain size on a logarithmic abscissa scale and voltage or efficiency on the ordinate. It was observed that a huge "valley" existed. For grain sizes between 10 nm and 1 millimeter, no good correlation could be observed between grain size and cell voltage or efficiency. In the 12 years following such plot, despite new experimental trials, not many new observations were added to Werner's original plot. This poses the question to what degree grain size could be an effective "driver" towards higher solar cell efficiency?

For many years, NREL had worked with the Astropower Corporation (Delaware, and its successor, GE) on developing thin crystalline Si solar cells and modules. They delivered various cell and module prototypes. What was striking was that with about 30 micron thick absorbers, short circuit current densities (<28 mA/cm²)and QE responses were measured for such cells that were similar to champion light enhanced nanocrystalline nc-Si:H-cells were the absorber was only 1.5 to 2 microns thick. This poses the question to what degree the fall-off of the quantum efficiency red response is determined by incomplete carrier generation or by incomplete carrier collection or both? Thin Si PV is a perfect example demonstrating where reliance on the appropriate R&D assumptions will greatly affect the optimization efforts. If losses were due to incomplete carrier collection, one would gear optimization attempts towards reducing collection losses, while incomplete generation losses would be fixed by enhancing generation, typically by applying optical enhancement schemes. It is possible that measured QE responses are affected by both factors, while a majority of R&D efforts may have been conducted under the assumption to enhance generation in thin solar cells by researching optical enhancement techniques alone. It is currently not known what the potential of crystalline Si film solar cells is. The observations made during the last 30 years optimizing Si based solar cells could suggest that progress with thin Si film solar cells could be less likely and may not be attainable, if the observations rather than the expectations were correct.

A similar question should be asked for another case of crystalline Si PV, recrystallized amorphous silicon. Commercial development for crystalline Si film solar modules has occurred at Pacific Solar (later CSG, Crystalline Si on Glass, at one point in time, affiliated with Q-Cells, now belonging to Suntech Corporation). CSG modules achieved about the same performance level as amorphous silicon (a-Si:H) modules. CSG uses 2 micrometer thick layers for the absorber (a recrystallized a-Si precursor on a specialty glass substrate). This observation poses a very fundamental R&D question: "Was CSG not given enough resources to develop a better solar cell, or was the expectation erroneous that a better solar cell efficiencies would result from recrystallized 2 micrometer-thick crystalline silicon layers (recrystallized 2 micron thick amorphous silicon) than using 0.5 micron thick a-Si:H layers directly to produce a solar cell or module?" In order for fundamental science to impact technology, a more conclusive answer to this question has to be found.

6. Relating champion cell efficiencies and commercial module performance

Champion efficiencies are often used as the yardstick to gauge the PV status of a certain technology. It was found that the credibility of such numbers improves when champion results of independent testing laboratories are used, although several PV entities have developed internal procedures to obtain, within experimental uncertainty, the same results as independent testing laboratories. What is more controversial is that sometimes "unoptimized" solar cell results have been reported. In those instances, it is not clear what efficiency level might be attainable upon further cell optimization. As argued in the introduction section of this chapter and elsewhere, it is not clear if greater control and reduction in variability increases or decreases champion cell efficiencies. It is recommended that for the time being, either possibility should be considered as likely, namely that unoptimized device performance can or cannot be further improved after full optimization.

Champion solar cell efficiencies can be linked to current and future commercial module performance, based on what is known today about solar cell champion efficiency levels, which in recent years have not shown too much progress. In order to obtain current

Eff. (%)	Module	T.coeff. (power)	Technology	Current c/c performance ratio (module/cell eff.)
19.5	SunPower E19/318	-0.38 %/C	mono-Si, special junction, sp. j. (1)	78% (19.5/25.0)
17.1	Sanyo HIP-215N	-0.34 %/C	CZ-Si, "HIT," sp. J (1)	69% (17.1/25.0)
15.1	Suniva ART245-60	-0.46%/C	CZ-Si, sp. J. (2)	72% (15.1/21.0)
14.3	Kyocera KD235GX-LPB	-0.44%/C	MC-Si, standard junction (std. j.)	70% (14.3/20.4)
14.3	Solar World SW235/240	-0.45%/C	CZ-Si, std. j.	68% (14.3/21)
14.3	Solar World SW 220/240	-0.48%/C	MC-Si, std. j.	72% (14.3/20.4)
13.9	Solaria 230/210	-0.5 %/C	"standard" mono-Si cells, 2x concentration	65%* (13.9/21.5)
13.6	Suntech STP 225-20Wd	-.0.44%/C	MC or CZ-Si, std. j.	67% (13.6/20.4)
13.6	Evergreen Solar ES 195	-0.49%/C	String-ribbon-Si std. j.	65% **(13.6/20.4)
12.5	Q-Cells Q.smart UF 95	-(0.38 %+/-0.04)%/C	CIGS	62% (12.5/20.3)
11.5	First Solar FS-382	-0.25%/C	CdTe	69% (11.5/16.7)
11.9	Avancis 130 W	-0.45%/C	CIGS	59% (11.9/20.3)
10.1	Abound Solar AB62/72	-0.37 %/C	CdTe	60% (10.1/16.7)
10.0	Sharp NA-NA-V142I I5/NA	-0.24%/C	a-Si/nc-Si	80% (10/12.5)
7.2	Uni-Solar PVL144	-0.21 %/C	a-Si, triple junction	60% (7.2/12.1)
6.3	Kaneka T-EC-120	n/a	a-Si single junction	62% (6.3/10.1)
1.7	Konarka Power Plastic 1140	+0.05%/C	organic	20% (1.7/8.3)

*There is no good published value for 2x concentrated cell performance. Here, the corresponding Solar cell efficiency is taken as 21.5%.
** There is some uncertainty whether or not string-ribbon Si can reach multicrystalline Si efficiencies, but this has been assumed.

Table 1. Module Efficiency from survey of manufacturers' websites and commercial module efficiency over champion cell efficiency ratios"

commercial performance, an internet survey provides some guidance as to what module products, and technologies are commercially available. Then the ratio between verified champion efficiency and module performance can be calculated. It is clear that commercially available module efficiencies have to be discounted from champion cell level efficiencies, and that module efficiencies are smaller than solar cell efficiencies. In 2006, this author used a "discount" of 20% between champion cells and commercial modules (von Roedern, 2006). Now, 5 years later, the data suggest that it would be very unlikely that average commercial module efficiencies could exceed 80% of the respective champion level solar cell efficiency. The most mature and selective technologies (wafer-Si) have not yet exceeded this (80%) value for even for their best commercial modules yet.

Table 1 shows a summary from February 2011 of commercially available PV modules. Only modules available on manufacturer's public websites for sale where the technology is identifiable are listed.

Using the 80% argument, it can then be estimated what maximum average commercial module efficiency is likely based on what is known about champion solar cells today. Table 2 provides such breakdown.

The point to be made is that there is a difference between champion cell and champion module efficiency (estimated to constitute an efficiency difference of about 20%). While some technologies may reach a high ratio earlier than others, current champion-cell efficiency numbers can be used to estimate future commercial module efficiencies. There are some claims that some modules perform better in hot environments than at low temperature. These real effects are on the order of +/- 10% in energy generation, but there are unknowns of similar magnitude like the degradation encountered over the system lifetime, the quality of the installation, the weather fluctuations, and the accuracy of the name-plate rating, to name a few factors.

Technology	Future commercial module performance (80% of current record cell efficiency)	Future Relative Performance	Future Relative-cost (using a 50% thin film cost advantage)
Silicon (non-stand)	19.8%	1.21	0.83(competitive)
Silicon (standard)	16.4%**	1.00	1.00 (reference)
Silicon (standard, 2x)	17.2%	1.05	0.71 (competitive)
CIS	16.2%	0.99	0.51 (highly competitive)
CdTe	13.2%	0.80	0.63 (highly competitive)
a-Si (1-j)	8.0%	0.49	1.02 (about the same)
a-Si (3-jj), (or a-Si/nc-Si)	9.8%	0.60	0.83 (competitive)

**Since there is only a marginal performance difference for standard cells using mono- or multi-Si wafers, an "average" champion cell performance of 20.5% was used to calculate standard Si module performance. 12.3% was used for spectrum splitting a-SiGe:H and nc-Si:H multijunctions

Table 2. Anticipated Future Module Efficiency and Relative Cost Based on Today's Demonstrated Champion Cell Performance

Low light-level efficiency values may look better for some modules then for others, but those higher low-light-level efficiencies can be lost after a module is deployed or stressed (Wohlgemuth 2010). Low light-level higher efficiency also affects the energy output differently in different climates. The more overcast the weather, the more important is lasting higher low light-level efficiency. The interactions between climate and energy output poses the question whether PV modules will get a single rating or a deployment site specific rating because modules are sold in STC Watts and revenues are received in terms of energy generated.

7. Notes on reliability and durability of thin-film modules

One of the most frequently questions asked is: How durable is this technology versus longer-established wafer-Si PV technology and whether or not Si PV would be "the hallmark of stability" as these technologies are sometimes presented. Clearly, there are changes in all technologies, and as modules age or are stressed, transient behavior, power and temperature coefficients will change (Whitaker et al., 1991, del Cueto & von Roedern, 2006, Wohlgemuth, 2010). I sense some reluctance in the testing community to specify accelerated stress conditions, because not all effects and mechanisms for module degradation or failure are known, and because some people in that community hold out the hope that there would be better accelerated stress conditions that would better predict real world performance of modules. What is not realized is that there could also be a value for having standards, and that standard conditions will not reflect real-world conditions or energy generation. For example, the fuel economy of automobiles is based on standard tests, while the prudent driver will know that he or she may not achieve or exceed the standards because of their driving techniques and conditions differ. I advocate that it is the manufacturer's duty to assure durability, and that long-term durability depends to a large degree on whether or not an appropriate manufacturing process was used. Technology related instability problems with any PV technology are currently difficult to identify, and mistakes were made in all technologies leading to the observation of unacceptably high module failure rates. Newer technologies are apt to reveal greater failure rates for a while.

While glass to glass sealed modules are often being produced, glass breakage can lead to increased failures in thin film PV modules. Glass breakage and its mitigation are the topics of much research. Standards are sometimes helpful and sometimes misleading. For example, two sheets of annealed glass laminated with a layer of ethyl vinyl acetate (EVA) can pass the hailstone impact test (a 2.5 cm diameter hailstone impacting at terminal velocity, 23 m/s). This has led many module developers assume that if the hail test can be passed, the mechanical strength was acceptable. Yet, thermally induced glass breakage will occur. Given the observation that glass breakage is not so much a factor for crystalline Si PV modules, the use of partially strengthened or partially tempered glass is strongly recommend to be used for thin-film PV modules.

It is also well established that some forms of PV are much more moisture or oxygen sensitive than other technologies. Sensitivity to ingress of elements can be mitigated by either making the device (solar cell) less sensitive to the penetrating elements, or by better sealing the module package. In practice, both approaches may be used to result in the most cost-effective scheme to increase the durability of a PV module. Glass will not allow penetration of elements and provides a perfect seal, except for the edge glass to glass seal. This may not be the case for flexible schemes where flexible layers have to be used. While flexible opaque materials may provide neccessay low transmission rates and do not pose a glass breakage problem, other issues may become critical when there is a need to use optically transparent barrier foils for flexible PV modules. An early example that the quality of a "package" needs to accommodate the sensitivity of the device was provided when the tested packaging approach used by industry (for a-Si:H-based PV module technology) did not sufficiently protect flexible CIGS modules. In fact, it was determined that the established (a-Si:H) package should be labeled as 'breathable,' as water vapor diffuses rather quickly through Tefzel and EVA. If the device can endure such water vapor transmission, its stability may be acceptable. There are now pilot-quantity flexible transparent barrier materials for niche applications for more sensitive PV technologies (like CIGS) with much lower (and perhaps adequately low) water transmission rates. What is more difficult to evaluate is how those materials compare in terms of cost to glass.

Most systems today are assessed by their energy output. That adds a complication, because in some climates modules with cracked glass may continue to perform well for a number of months or even years. Because glass breakage is very evident, and because these broken modules are not likely to deliver guaranteed powers after many years and may present a safety problem, broken modules may get replaced before they cause a notable power loss. Similar arguments apply to the effect of delamination. If modules get replaced as soon as a visible defect appears, it may become more difficult to assess average long-term stability. An added problem is that it is hard to predict how delamination will progress. One thing to notice is that T-coefficients for power may become smaller negative (or for stabilized a-Si:H- or OPV-based PV even slightly positive) numbers as the modules are being deployed. Smaller than wafer Si PV negative temperature coefficients are typically viewed as something positive, as the derate going from an STC to a real world condition rating decreases. However, if the T-coefficient were to become a less negative number upon deployment, one has to keep in mind that the STC degradation may actually increase more rapidly than the outdoor data might suggest.

The testing community is looking to develop rapid tests that can reliably predict long-term module performance. Such development requires an understanding about all major mechanisms leading to long-term power loss. Only after individual mechanisms are known can there be an assessment how they will respond to acceleration. Then, perhaps more appropriate tests could be developed. In the mean time, much "infant mortality" of PV modules can be avoided by passing qualification tests. For example, when the "wet high potential test" (wet high pot) test was being implemented, modules having defects in the edge seal were identified and eliminated. While the wet high pot test was originally conceived out of safety concerns, it was also useful for eliminating early module failures. Further testing of leakage currents is important, and modules should perhaps be tested not only to the safety standard but rather to the lowest leakage current that can be measured for a specific module configuration. For wafer Si PV modules, much progress with respect to module durability was achieved by passing the JPL "block" tests that later resulted in the appropriate qualification test (e.g., IEC 61215, 61730). However, one should not forget that a module passing qualifications tests may fall below guaranteed (warranted) power in the field while modules that could not pass qualification tests may show acceptable durability upon long-term deployment (Wohlgemuth et al., 2006).

Further (beyond not understanding all mechanisms in detail), the accurate prediction of lifetime details is further encumbered by the statistical nature of the degradation behavior, leading to a spread in the observed data. Hence, rather than testing individual modules, statistically relevant identical module samples have to be assessed. The other issue is that outdoor conditions vary and cannot be in detail predicted. The latter observation poses the question whether module manufacturers will develop modules for specific climates, or whether there will be one product for all climates. Whether or not we will see differentiation in the modules for weather-specific sites will undoubtedly depend on the cost savings encountered if/when climate-specific modules are manufactured. Many industrial items, say automobiles or consumer electronics, are manufactured such that only a single quality standard and product exists. Customers like 'rankings' of items using standardized procedures or tests but do often not realize that if the difference between ranks is less than the uncertainty there may be statistically no difference between those ranks.

There cannot be absolute certainty about the warrantee period until such time has passed. Typical wafer Si PV guarantees given about 20 years ago correctly predicted that such modules or PV arrays would provide on average 80% or more of their initial rating. Today, typically such power warrantee increased to 80% of minimum rated power output after 25 years. Manufactures give 'competitive' warranties, which in addition to technical reasons define the typical 25-year power warranty period. Since wafer Si PV is providing such guarantees, the competing thin-film PV companies have to do so as well. In the opinion of the author, such warranties will likely be met by many reputable manufacturers. However, the numbers are quite staggering. If in 2010 about 15 GW of PV were sold world-wide and if 1.5 GW of modules installed in 2010 required replacement before 2035 due to low power (assumption: 10% of modules require warranty replacement because of more than the guaranteed power loss has occurred), that is 5 million 300-W modules, and corresponds to the wattage manufactured in 2010 by one of the world's largest PV companies. While no predictions can be made with absolute certainty, it is somewhat likely that all enduring PV modules, including thin-film PV module technologies, will meet or exceed current limited power warranty of 80% after 25 years.

8. Outlook

Future development of PV technologies is uncertain. Table 2 provided the author's current outlook on efficiency and relative costs. It is difficult to project real PV costs far enough into the future. However, Table 2 also shows that projections are possible based on what is known today about specific PV technologies. Table 2 also provides an example of why it is important to make independently verified champion solar cells. "Champion" solar cell efficiency numbers provide historic continuity, as they have served as a "yardstick" to progress within each PV technology. Looking at crystalline Si PV, it is not clear if standard or non-standard approaches will gain or lose market share. Table 2 essentially says that if the cost reduction is proportional to an efficiency decrease, there is no net economical benefit.

Whenever observations do not confirm expectations, it is suggested to question expectations with the same scrutiny as observation (experimental results). The statistical nature of data needs to be realized; it should be always said what is being compared, best, average or worst data. For solar cell efficiencies, this requires an understanding to distinguish between best (champion) and average production efficiencies. Sometimes, advantages and disadvantages of a process change are not pointed out with the same scrutiny. Researchers have to ask themselves whether there should be further optimization of known factors, or if greater progress could be made being guided by unexpected or empirical results. Historic examples exist for new results being developed guided by a flawed theory (e.g., the invention of black powder) or the guidance of a correct theory could lead to unexpected results (Columbus discovering America while searching for a new route to India). It is especially important to keep observations and already established results in mind to avoid unnecessary repetition of experiments. Without this, unfruitful approaches to solar cell development could be tried anew.

It is important to realize the role of material science in this process. On one hand, it is known that higher quality materials can result in higher solar cell performance, while on the other hand it is also known that sometimes the incorporation of "inferior" material layers resulted in champion level efficiency cells. The use of CBD CdS, resistive TCO, and polycrystalline, non-stoichiometric, Na-laden CIGS films on glass rather than single crystal CIGS makes that point. It is well known that solar cell optimization is "interactive," i.e., when one layer in a cell is improved, other layers may need to be reoptimized. For example, when the TCO layer in an a-Si:H-based solar cells were switched from SnO_2 to ZnO, the p-layer deposition conditions also had to be reoptimized to obtain the highest efficiency solar cell or module after such switch. A fundamental answer has to be found for the following question: Why is a high-lifetime mono-crystalline silicon wafer easily processed into a low efficiency solar cell? In addition, the following question requires an answer: "Is there a single set of parameters defining stabilized champion solar cells, or are multiple combinations of materials and solar cell parameters (V_{OC}, J_{SC}, and FF) capable of reaching champion level cell efficiencies? Recent observation in the case of CIGS solar cells suggests that there could be indeed multiple optima.

The proprietary nature sometimes hurts the development of correct models. For example, to correctly identify the stability mechanisms in solar cells or modules, all processing detail may have to be known. Often, companies do not wish to make such knowledge public. In these instances, it appears most effective to bring together researchers in a conference or workshop setting to discuss as much of a problem as is possible.

It is not clear which technologies will "win" in the long run. Thin films have a cost advantage over crystalline Si, provided the durability is comparable and the performance is high enough. Arguments were presented that the benefit from moving from wafer Si to thin film products can be calculated.

9. Acknowledgement

The author would like to thank the many colleagues without who's knowledge, capabilities, and expertise this chapter would not have been possible. This work was supported by the U.S. Department of Energy under Contract No. DE-AC36-08GO28308 with the National Renewable Energy Laboratory.

10. References

Anderson, T.J., Crisalle, O.D. Li, S.S. & Holloway, P.H. (2003). Future CIS Manufacturing Technology Development. NREL/SR-520-33997, see the entire report

Attar, G., Muthaiah, A., Natarajan, H., Karthikeyan, H., Zafar S., Ferekides, C. S. & Morel, D.L. (1994). Development of Manufacturable CIS Processing. *Proceedings of the First World Conference on Photovoltaic Energy*, (Waikoloa, HI, 5.-9.12.1994), pp. 182-185. ISBN 0-7803-1459-X

Catalano, A., D'Aiello, R.V., Dresner, J., Faughnan, B., Firester, A., Kane, J., Schade, H., Smith, Z.E., Swartz, G. & Triano, A. (1982). Attainment of 10% Conversion Efficiency in Amorphous Silicon Solar Cells *Conference Record of the 16th IEEE Photovoltaic Specialists Conference*, (San Diego, CA, 27-30.9.1982), pp. 1421-1422. ISSN 0160-8371

Chambouleyron, I. & Alvarez, F., (1985). Conversion Efficiency of Multiple-Gap Solar Cells under Different Irradiation Conditions. *Conference Record of the 18th IEEE Photovoltaic Specialists Conference*, (Las Vegas, NV, 21-25.10.1985), pp. 533-538. ISSN 0160-8371

Contreras, M.A., Nakada, T., Hongo, M., Pudov, A.O., & Sites, J.R., (2003). ZnO/ZnS(O,OH)/Cu(In,Ga)Se$_2$ Solar Cell with 18.6% Efficiency. *Proceedings of the. 3rd World Conference on Photovoltaic Energy Conversion*, Osaka Japan, paper 2LN-C-08

Cunningham, D.W., Frederick, Gittings, Grammond, Harrer, S., Intagliata, J., O'Connor, N.,Rubcich, M., Skinner, D., & Veluchamy, P., (2002). Progress in Apollo®

Technology. *Conference Record of the 29th IEEE Photovoltaic Specialists Conference*, (New Orleans, 5.19-24.2002), pp. 559-562. ISBN 0-7803-7471-1

Dobson, K., Visoly-Fisher, I., Hodes, G., and Cahen, D. (2000). Stability of CdTe/CdS thin-film solar cells. *Solar Energy Materials and Solar Cells*, 62 (2000) pp. 295-325. ISSN 0927-0248

del Cueto, J.A. & von Roedern, B. (2006). Long-term transient and metastable effects in cadmium telluride photovoltaic modules. *Progress in Photovoltaics: Research & Applications* 14, 615-628. ISNN 1099-159X, (an example for CdTe PV)

Enzenroth, R. A., Barth, K.L. & Sampath, W.S. (2005). Correlation of stability to varied $CdCl_2$ treatment and related defects in CdS/CdTe PV devices as measured by thermal admittance spectroscopy. *Journal of Physics and Chemistry of Solids*, 66 pp. 1883-1886. ISSN 0022-3697

Gabor, A.M., Tuttle, J., Albin, .D.S., Contreras, M.A., Noufi, R., & Hermann, A. M., (1994). High Efficiency $CuIn_xGa_{1-x}Se_2$ Solar Cells made from $In_xGa_{1-x})_2Se_2$ precursor films. *Applied Physic Letters* 65, pp. 198-200. ISSN 0003-6951

Green, M.A., Emery, K., Hishikawa, K.Y. & Warta, W. (2011). Solar Cell Efficiency Tables (version 37). *Progress in Photovoltaics: Research and Applications 19*, pp. 84-92. ISSN 1099-159X. In some instances, results from earlier such tables or results from the "notable exceptions" tables are used

Guha, S., Yang, J., Pawlikiewicz, A., Glatfelter, T., Ross, R. & Ovshinsky S.R. (1988). A Novel Design for Amorphous Silicon Solar Cells. *Conference Record of the 20th IEEE Photovoltaic Specialists Conference*, (Las Vegas, NV, 26-30.9.1988), pp. 79-84. ISSN 0160-8371

Izu, M., Deng, X., Krisko, A., Whelan, K., Young, R., Ovshinsky, H. C., Narasimhan, K. L. & Ovshinsky, S. R., (1993). Manufacturing of Triple-Junction 4 ft² a-Si Alloy PV Modules. *Conference Record of the 23rd IEEE Photovoltaic Specialists Conference*, (Louisville, KY, 10-14.5.1993), pp. 919-925. ISBN 0-7803-1220-1

Kuwano, Y., Ohniishi, Nishiwaki, H., Tsuda, S., Fukatsu, T., Enomoto, K., Nakashima, Y., and Tarui, H., (1982). Multi-Gap Amorphous Si Solar Cells Prepared by the Consecutive, Separated Reaction Chamber Method. *Conference Record of the 16th IEEE Photovoltaic Specialists Conference*, (San Diego, CA, 27-30.9.1982), pp. 1338-1343. ISSN 0160-8371

Lee, Y., Jiao, L. H., Liu, H., Lu, Z., Collins, R.W. & Wronski, C. R., (1996). Stability of a-Si :H Solar Cells and Corresponding Intrinsic Materials Fabricated Using Hydrogen Diluted Silane. *Conference Record of the 25th IEEE Photovoltaic Specialists Conference*, (Washington, DC, 13-17.5.1996), pp. 1165-1168. ISBN 0-7803-3166-4

Luft, W., Stafford, B., von Roedern, B., & DeBlasio, R. (1992). Preospects of amorphous silicon photovoltaics. *Solar Energy Materials and Solar Cells*, 26, pp. 17-26. ISSN 0927-0248

Luysberg, M., Scholten, C., Houben, L., Carius, R., Finger, F. & Vetter, O., (2001). Structural Properties of Microcrystalline Si Solar Cells. *Materials Research Society Symposia Proceedings* 664, pp. A15.2.1-6. ISBN 1-55899-600-1

McCandless, B. E., (2001). Thermochemical and Kinetic Aspects of Cadmium Telluride Solar Cell Processing. *Materials Research Society Symposia Proceedings* 668 (San Francisco, CA 16-20.4.2001), pp. H1.6.1-12. ISBN 1-55899-604-4

Noufi, R., (2010). Private communication

Platz, R., Pellaton Vaucher, N., Fischer, D., Meier, J. & Shah, A., (1997). Improved Micromorph Tandem Cell Performance through Enhanced Top Cell Currents. *Conference Record*

26th IEEE Photovoltaic Specialists Conference, (Anaheim, CA, 29.9-3.10.1997), pp. 691-694. ISBN 0-7803-3767-0

Schmid, D., Ruckh, M., Grunwald, F. & Schock, H.W. (1993). Chalcopyrite/defect chalcopyrite heterojunctions on the basis of CuInSe₂. *Journal of Applied Physics* 73, pp. 2902-2909. ISSN 0021-8979

Shah, A., Sauvain, E., Wyrsch, N., Curtins, H., Leutz, B., Shen, D. S., Chu, V., Wagner, S., Schade, H. & Chao, H. W. A. (1988). a-Si:H Films Deposited at High Rates in 'VHF' Silane Plasma : Potential for Low-Cost Solar Cells. *Conference Record of the 20th IEEE Photovoltaic Specialists Conference,* (Las Vegas, NV 26-30.9.1988), pp. 282-287. ISSN 0160-8371

Tarrant, D.E. & Gay, R. R., (1995). Research on High-Efficiency, Large-Area CuInSe2-Based Thin-Film Modules, NREL/TP-413-8121. The fabrication sequence of Siemens Solar is shown on pages 2 & 3 in Figures 2 & 3

Todorov, T. K., Reuter, K.B. & Mitzi, (2010). High Efficiency Solar Cell with Earth-Abundant Liquid-Processed Absorber. *Advanced Materials* 22, pp. E156-159. ISSN 1121-4095.

Townsend, S.W., Ohno, T.R., Kaydanov, V., Gilmore, A.S., Beach, J.D. & Collins, R.T. (2001). The Influence of Stressing at Different Biases on the Electrical and Optical Properties of CdS/CdTe Solar Cells.. *Materials Research Society Symposia Proceedings* 668 (San Francisco, CA 16-20.4.2001), pp. H5.11.1-6. ISBN 1-55899-604-4

von Roedern, B. & del Cueto, J.A. (2000). Model for Staebler-Wronski Degradation Deduced from Long-Term, Controlled Light-Soaking Experiments. *Mataterials Research Society Symposia Proceedings* 609, (San Francisco, CA24-28.4.2000), pp. A10.4.1-6. ISBN 1-55899-517-X

Wanlass, M.W. & Albin, D.S., (2004). A Rigerous Analysis of Series-Connected, Multi-Bandgap, Tandem Thermophotovoltaics (TPV) Energy Converters. *Proceedings of the 6th Conference on Thermophotovoltaic Generation of Electricity,* (Freiburg, Germany 14-16.6.2004), American *Institute of Physics Conference Proceedings* 738, pp. 462-470. ISBN 0-7354-02221

Whitaker, C.M., Townsend, T. U., Wenger, H. J., Illiceto, A., Chimento, G. & Paletta, F. (1991). Effects of Irradiance and Other Factors on PV Temperature Coefficients. *Conference Record of the 22nd IEEE Photovoltaic Specialists Conference,* (Las Vegas, NV 7-10.10.1991), pp. 608-613. ISBN 0-87942-635-7

Yan, B., Yue, G. & Guha, S., (2007). Status of nc-Si :H Solar Cells at United Solar and Roadmap for Manufacturing a-Si :H and nc-Si :H Based Solar Panels. *Materials Researchy Society Symposia Proceedings* 989 (San Francisco, CA, 9-13.4.2007), pp. 335-346. ISBN 978-1-55899-949-7

von Roedern, B. (2006). Thin Film PV Module Review. *Refocus magazine (Elsevier) (July + August 2006)* pp. 34-36. ISSN 1471-0846

Wohlgemuth, J.H. (2010). private communication

Wohlgemuth, J.H., Cunningham, D.W., Monus, P., Miller, J. & Nguyen, A., (2006). Long- term reliabilty of photovoltaic modules. *Conference Record of the 2006 IEEE 4th World Conference on Photovoltaic Energy Conversion* (Waikoloa, HI 7-12.5.2006), pp. 2050-2053. ISBN 1-4244-0017-1

11

Spectral Effects on CIS Modules While Deployed Outdoors

Michael Simon and Edson L. Meyer
Fort Hare Institute of Technology, University of Fort Hare
South Africa

1. Introduction

The effect of spectral distribution on the performance of photovoltaic (PV) modules is often neglected. The introduction of multi-junction devices such as Copper Indium Diselenide (CIS) necessitated a concerted investigation into the spectral response on these devices. In part this attributed to the wider spectral response resulting from a combination of different energy band gaps. This in turn implies that the device should have a relatively lower dependence on outdoor spectral content, which depends on a number of factors such as year time, location, day time and material composition in the atmosphere.

The availability of outdoor spectral data, which in most cases is not available, allows for the evaluation of the outdoor response of the CIS technology as the spectrum shifts during the course of the day, during cloud/clear sky condition and seasons. This study reports on the effect of outdoor spectrum, which is different from the reference AM 1.5, on the CIS performance parameters.

2. Different outdoor methodologies currently adopted

2.1 The concept of average photon energy

In trying to quantify the 'blueness' or 'redness' of outdoor spectrum, Christian *et. al.* adopted the concept of Average Photon Energy (APE) as an alternative (Christian et al., 2002). He defined APE as a measure of the average hue of incident radiation which is calculated using the spectral irradiance data divided by the integrated photon flux density, as in equation 1.

$$APE = \frac{\int_a^b E_i(\lambda)d\lambda}{q_e\int_a^b \Phi_i(\lambda)d\lambda} \qquad (1)$$

where :

	q_e	=	electronic charge
	$E_i(\lambda)$	=	Spectral irradiance
	$\Phi_i(\lambda)$	=	Photon flux density

As an indication of the spectral content, high values of average APE indicate a blue-shifted spectrum, whilst low values correspond to red shifted spectrum. Although this concept at first approximation characterizes the spectral content at a particular time-of-the day, no direct feedback of the device information is obtained since it is independent of the device. The concept of Average Photon Energy (APE) has also been adopted to illustrate the seasonal variation of PV devices (Minemoto et al., 2002; Christian et al., 2002).

2.2 The Air Mass concept

The mostly commonly adopted procedure (Meyer, 2002; King et al., 1997) is to calculate the Air Mass (AM) value at a specific location and relate the module's electrical parameters. It is standard procedure for PV manufacturers to rate the module's power at a specific spectral condition, AM 1.5 which is intended to be representative of most indoor laboratories and is not a typical spectral condition of most outdoor sites. The question that one has to ask is, why then is AM 1.5 spectrum not ideal? What conditions were optimized in the modeling of AM 1.5 spectra? What are the cost implications on the customer's side when the PV module is finally deployed at spectra different from AM 1.5?

The modeled AM 1.5 spectrum commonly used for PV module rating was created using a radiative transfer model called BRITE (Riordan et al., 1990). The modeled conditions used for example the sun-facing angle, tilted 37° from the horizontal, was chosen as average latitude for the United States of America. The 1.42 cm of precipitable water vapor and 0.34 cm of ozone in a vertical column from sea level are all gathered from USA data. Ground reflectance was fixed at 0.2, a typical value for dry and bare soil. In principle this spectra is a typical USA spectrum and therefore makes sense to rate PV modules which are to be deployed in USA and the surrounding countries.

AM is simply defined as the ratio of atmospheric mass in the actual observer - sun path to the mass that would exist if the sun was directly overhead at sea level using standard barometric pressure (Meyer, 2002). Although the concept of AM is a good approximation tool for quantifying the degree of 'redness' or 'blueness' of the spectrum, the major draw back is that it is applied under specific weather conditions, i.e., clear sky, which probably is suitable for deserts conditions.

2.3 The spectral factor concept

Another notion also adopted to evaluate the effect of outdoor spectrum, is the concept of Spectral Factor. As described by Poissant (Poissant et al., 2006), Spectral Factor is defined as a coefficient of the short-circuit current (I_{sc}) at the current spectrum to the short-circuit current at STC (I_{STC}).

$$m_t = \frac{I_{sc}}{I_{STC}} \cdot \frac{\int_{\lambda_1}^{\lambda_2} E_{STC}(\lambda)d\lambda}{\int_{\lambda_1}^{\lambda_2} E(\lambda)d\lambda} \tag{2}$$

From equation 2, the I_{sc} and the I_{STC} is obtained using the equation 3 and 4 respectively.

$$I_{sc} = \int_{\lambda_1}^{\lambda_2} E(\lambda)R_t(\lambda)d\lambda \tag{3}$$

$$I_{STC} = \int_{\lambda_1}^{\lambda_2} E_{STC}(\lambda)R_t(\lambda)d\lambda \tag{4}$$

where: $E(\lambda)$ = Irradiance as function of wavelength
$E_{STC}(\lambda)$ = Irradiance at STC
$R(\lambda)$ = Reflectivity

The spectral factor quantifies the degree of how the solar spectrum matches the cell spectral response at any given time as compared to the AM1.5 spectrum.

2.4 The useful fraction concept

With regard to changes in the device parameters, the concept of Useful Fraction used by Gottschalg et al (Gottschalg et al., 2003) clearly demonstrates the effect of varying outdoor spectrum. Useful fraction is defined as the ratio of the irradiance within the spectrally useful range of the device to the total irradiance.

$$UF = \frac{1}{G} \int_0^{E_g} G(\lambda)S(\lambda)d\lambda \tag{5}$$

Where E_g is the band-gap of the device (normally the cut - off wavelength) and G is the total irradiance determined as:

$$G(\lambda) = \int_0^{\lambda_{cut-off}} G(\lambda)d\lambda \tag{6}$$

where $G(\lambda)$ is the spectral irradiance encountered by a PV cell.

3. Methodology used in this study

Before the CIS module was deployed outdoors, the module underwent a series of testing procedures in order to establish the baseline characteristics. Visual inspection was adopted to check for some physical defects e.g. cracks, and incomplete scribes due to manufacturing errors. Infrared thermography revealed that no hot spots were present before and after outdoor exposure. These procedures were used to isolate the spectral effects with respect to the performance parameters of the module. To establish the seasonal effects on the module's I-V curves, three I-V curves were selected. One I-V curve for a winter season and the 2nd I-V curve for summer season were measured. The 3rd I-V curve was used to establish whether the module did not degrade when the winter curve was measured. All curves were measured at noon on clear days so that the effect of cloud cover would be negligible. For accurate comparison purposes all I-V curves had to be normalized to STC conditions so that the variations in irradiance and temperature would be corrected. Firstly the I_{sc} values were STC corrected by using equation 1 (Gottschalg et al., 2005).

$$I_{sc} = \left(\frac{I_{sc}}{G} \times 100\right) + \left(25 - T_{module}\right) \times \alpha \tag{7}$$

where a is the module temperature coefficient [A/°C].

Each point on the I-V curve had to be adjusted according to equation 8.

$$I_2 = I_1 + I_{sc} \times \left[\left(\frac{1000}{G} \right) - 1 \right] + \alpha \left(25 - T_{module} \right) \tag{8}$$

where: I_1 =   measured current at any point
              I_2 =   new corrected current
              G =   measured irradiance

The corresponding voltage points were also corrected according to equation 9.

$$V_2 = V_1 - R_s \times \left(I_2 - I_1 \right) + \beta \times \left(25 - T_{module} \right) \tag{9}$$

where: V_1 =   measured voltage at a corresponding point for I_1
              R_s =   internal series resistance of the module [Ω]
              β =   voltage temperature coefficient of the module [V/$\circ$C]
              V_2 =   new corrected voltage

The outdoor spectrum was also measured for winter and summer periods in order to compare them for possible changes in the quality of the two spectra (figure 5). With regard to changes in the device parameters, the concept of Weighted Useful Fraction (WUF) (Simon and Meyer, 2008; Simon and Meyer, 2010) was used to clearly demonstrate the effect of varying outdoor spectrum. This concept was developed due to some limitations noted with other outdoor spectral characterization techniques (Christian et al., 2002).

The methodology used by Gottschalg et al (Gottschalg et al, 2002) makes use the assumption that the energy density (W/m²/nm) within the spectral range of the device at a specific wavelength is totally absorbed (100%). But in reality the energy density at a specific wavelength has a specific absorption percentage, which should be considered when determining the spectral response within the device range. It was therefore necessary to introduce what is referred to as the Weighted Useful Fraction (WUF) (Simon and Meyer, 2008; Simon and Meyer, 2010).

$$WUF = \frac{1}{Gtot} \int_0^{E_g} G(\lambda) d(\lambda) SR(\lambda) \tag{10}$$

where: $G(\lambda)$ is the integrated energy density within device spectral range with its corresponding absorption percentage evaluated at each wavelength.

As a quick example, at 350 nm for a-Si device, its corresponding energy density (W/m²/nm) is 20% of the irradiance (W/m²) received which contribute to the electron-hole (e-h) creation and for mc-Si at the same wavelength, 60% is used to create e-h pairs. But the concept of Useful Fraction considers that at each wavelength, all the energy received contributes to the e-h, which is one of the short comings observed from this methodology. The idea of using Weighted Useful Faction was to address these short falls which tend to over estimate the overall device spectral response.

The data obtained using the concept of Weighted Useful Fraction represents a statistical phenomenon of occurrences. Therefore the Gaussian distribution as a statistical tool was used to interpret the data simply because of a mathematical relationship (Central Limit Theorem). In this case the theorem holds because the sample is large (major condition of the theorem) and therefore the Gaussian distribution is suitable to be applied. In this study, the

3rd parameter Gaussian distribution function was used to describe the distribution pattern and to accurately determine the variance of points from the peak value (central value). The peaks of the Gaussian distribution was obtained by firstly creating frequency bins for the WUF and determine the frequency of the points in each bin expressed as a percentage. The bins were imported into SigmaPlot 10 and the peak 3rd Gaussian distribution function was used to accurately generate the peak WUF. Figure 1 illustrates the frequency distribution bins for a-Si:H module.

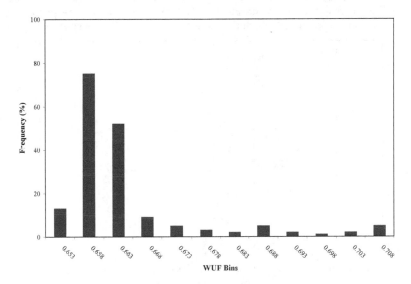

Fig. 1. Frequency distribution of WUF for a-Si:H module

Evident from figure 1 is an increase in WUF frequency at specific WUF value. This percentage frequency represents the number of data points measured at a specific WUF during the study period.

The centre of the points, which corresponds to the spectrum the device "prefers" most, was obtained using the peak Gaussian distribution of the form:

$$f = a\exp\left[-0.5\left((x-x_o)/b\right)^2\right] \tag{11}$$

where: a = highest frequency
                  x = WUF value
                  x_0 = WUF centre value
                  b = deviation (2σ)

Figure 2 illustrates a typical Gaussian distribution used to accurately determine the mean Weighted Useful Fraction.

Also illustrated is the width of the distribution as measured by the standard deviation or variance (standard deviation squared = σ^2). In order to interpret the results generated from each Gaussian distribution, two main terminologies had to be fully understood so that the results have a physical meaning and not just a statistical meaning. The standard deviation (σ) quantifies the degree of data scatter from one another, usually it is from the mean value.

In simple statistics, the data represented by the Gaussian distribution implies that 68% of the values (on either side) lie within the 1st standard deviation (1σ) and 95% of the values lie within the 2nd standard deviation. The confidence interval level was also analyzed when determining the mean value. The confidence interval quantifies the precision of the mean, which was vital in this analysis since the mean represents the WUF spectrum from which the devices responds best during the entire period of outdoor exposure. The increase in standard deviation means that the device spends less time on the corresponding WUF spectrum. Ideally it represents the error margin from the mean value. The percentage frequency value corresponding to the mean WUF value represents the percentage of the total time of outdoor exposure to which the device was responding best to that spectrum.

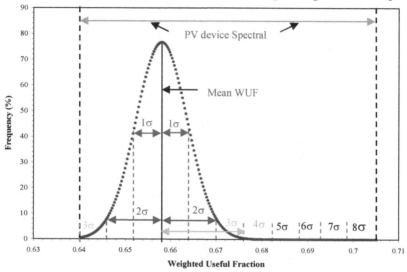

Fig. 2. Illustration of Gaussian distribution used to determine the mean WUF.

Depending on how the data is distributed, the Gaussian curve 'tails' differently from each side of the mean value. The increase in σ in this case reveals two crucial points regarding the statistical data in question. Firstly, it quantifies the total time spent at a specific spectrum as the σ increases during the entire period of monitoring. Secondly it reveals the entire spectral range to which PV devices respond. From figure 2, the standard deviation increases from 1σ to 8σ on one side of the mean WUF and from the other side varies from 1σ to 3σ. The total range of the WUF is from 0.64 to 0.7 although it spends less time from spectral range where standard deviation σ is greater than a unit. A high confidence level of each Gaussian distribution indicates the accuracy of the determined mean. All results presented in this work showed a high confidence level.

Normalization of I_{sc} was achieved by dividing the module's I_{sc} with the total irradiance within the device spectral range ($G_{Spectral\ Range}$). The commonly adopted correlation existing between the module's I_{sc} and back-of-module temperature is of the form $I_{sc} = (C_0 + C_1 T_{device}) \times G_{SpectralRange}$ (Gottschalg et al., 2004). Firstly, the relationship between $I_{sc} \big/ G_{SpectralRange}$ (which is referred to as $\varphi_{SpectralRange}$ from this point onwards) is plotted against back-of-module temperature. The empirical coefficient C_0 and C_1 are obtained. The second aspect is to plot $\varphi_{SpectralRange} \div (C_o + C_1 T_{device}) = f(WUF)$ versus the Weighted Useful Fraction (WUF), from which the predominant effect of the spectrum can be observed and analyzed. Due to a large number of data obtained, all results analyses were made using only data corresponding to global irradiance (G_{global}) > 100 W/m². This was done to reduce scatter without compromising the validity of the results

4. Results and discussion

Although the outdoor parameters might 'mimic' the STC conditions, the performance of the PV device will not perform to that expectation. By analyzing the effect of outdoor environment, the spectrum received is largely influenced by solar altitude and atmospheric composition, which in turn affect device performance.

Figure 3 illustrates the seasonal effects on the CIS module current-voltage *(I-V)* characteristics when deployed outdoor, first on 31 January 2008 and later on 12 June 2008.

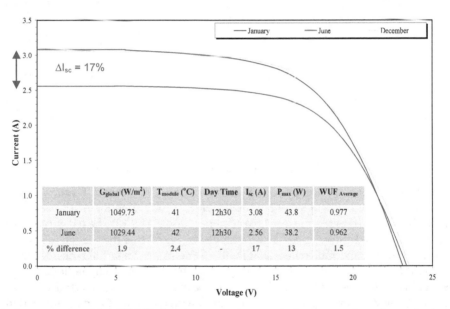

	G_{global} (W/m²)	T_{module} (°C)	Day Time	I_{sc} (A)	P_{max} (W)	WUF $_{Average}$
January	1049.73	41	12h30	3.08	43.8	0.977
June	1029.44	42	12h30	2.56	38.2	0.962
% difference	1.9	2.4	-	17	13	1.5

Fig. 3. Comparison of the CIS I-V characteristics for a typical summer clear sky and winter clear sky. The accompanying table lists the conditions before corrections to STC.

The January *I-V* curve was taken a few days after deployment of the modules while operating at outdoor conditions. Two aspects needed to be verified with this comparative analysis of the I-V curves for that time frame: Firstly the state of the module, i.e. whether it did not degrade within this time frame needed to be ascertained so that any effect on device I_{sc}, FF and efficiency would be purely attributed to spectral effects. Secondly, this was done to see the effect of seasonal changes on the *I-V* characteristics. Since the outdoor conditions are almost the same when the measurements were taken, the I-V curves were normalized to STC conditions using the procedure mentioned in section 2. Since the 3 *I-V* curves had been corrected for both temperature and irradiance, therefore any modification or changes on the I_{sc} values is purely due to spectral effect. The difference in module's I_{sc} is largely attributed to the outdoor spectral composition, which as have been mentioned earlier on, depends on season and time of the year amongst other factors. The CIS module was also simulated using Solar Studio Design. At each AM value, the module's I-V curve was obtained. Figure 4 illustrates the effects on the simulated CIS I-V curves as the Air Mass was varied.

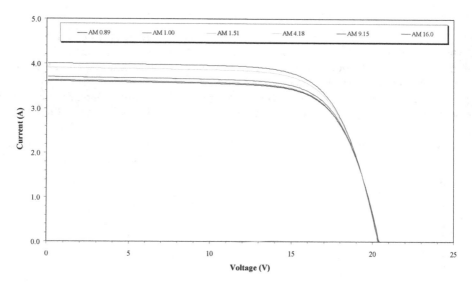

Fig. 4. The effect of varying Air mass on the simulated CIS module.

The change in outdoor spectrum as characterized by the AM values affect the module's I-V curves, mostly the I_{sc}. Although this module is rated at STC using the AM1.5 spectrum, the CIS module is performing less at AM1.5 as compared to AM 9.15. The I-V curve at AM 1.5 coincides with the I-V curve at AM 16.0. It should be noted that the change in AM value is an indication of the spectral content dominating. The ΔI_{sc} = 7.5% difference between I_{sc} at AM 1.5 and I_{sc} at AM 9.15 is purely due to spectral changes. Returning back to figure 1, the difference in I_{sc} between winter and summer spectrum is due to spectral changes. The typical winter and summer spectra were compared with the view of finding any variation in the profiles. All values were divided by the highest energy density in each curve so as to normalize them. Figure 5 presents the normalized spectral distribution corresponding to the two I-V curves in figure 3.

Clearly there is a difference in the spectral content primarily due to the difference in solar altitude and hence air mass. In the absence of the device degradation, similar irradiance and module temperatures, the reduction in module performance is attributed to the difference in spectral distribution associated with the seasonal variation. To further verify whether indeed the reduction in the module's I_{sc} was due to spectral changes associated with seasonal changes, the device WUF for the entire year was analyzed. The monthly average WUF was considered to be sufficient to provide evidence, if any in its profile. Figure 6 shows the evolution of the monthly average WUF of the CIS module.

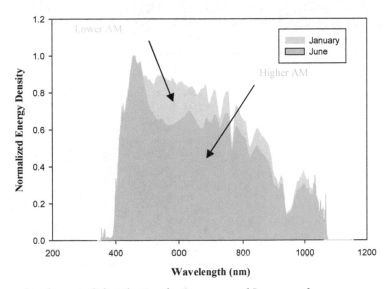

Fig. 5. Normalized spectral distribution for January and June months.

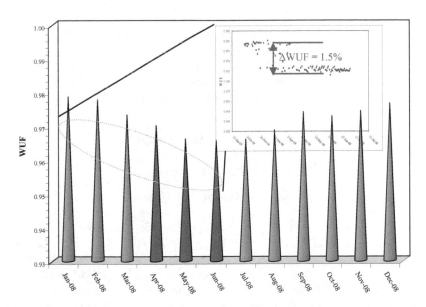

Fig. 6. Evolution of daily average Weighted Useful Fraction versus timeline. Inset is an average daily profile for the period from January to June 2008.

Evident from figure 6 is the high values of CIS WUF for the entire period which indicates that the device performs well under full spectrum. Taking the average values of the upper (summer) and the lower for winter, a 1.5% drop in WUF is noticed (inset figure). A small change in WUF results in large change of the device's I_{sc}. In order to verify this assumption, the change in WUF versus Air Mass was established as is presented in figure 7.

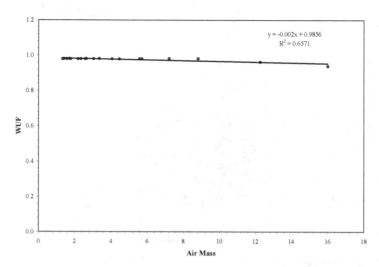

Fig. 7. Influence of the air mass on device spectral variations as characterized by WUF for CIS module.

The relationship established in figure 7 was used to calculate the change in WUF at different Air Mass values, a typical change in season. Values for low air mass (indication of a summer spectrum) and high air mass (indication of winter) were used to calculate the % change in WUF and later compared to the simulated % change in I_{sc} at different AM values, the same values that has been used in previous calculation. Equations 12 and 13 illustrate the equations used for this calculation.

$$WUF_{1.0} = -0.002 \times AM1.0 + 0.9856 \tag{12}$$

$$WUF_{9.15} = -0.002 \times AM9.15 + 0.9856 \tag{13}$$

where: $WUF_{1.0}$ is the calculated value of WUF at AM 1.0 and the $WUF_{9.15}$ is the calculated value of WUF at AM 9.15.

From figure 4 the value for I_{sc} (AM 1.0) and I_{sc} (AM 9.15) were used to calculate the % change in I_{sc} as the spectrum changes. The $\Delta WUF = WUF_{1.0} - WUF_{9.15}$ expressed as a %, was found to be 1.66%, while the $\Delta I_{sc} = 11.88\%$. From this analysis, one can conclude that a small % change in ΔWUF result in large % difference of the module's I_{sc}, which explains the 17% decrease in I_{sc} due to a ΔWUF of 1.5%. The slight difference in the two results is due to the difference in the actual operating conditions in which case the simulated conditions are different from the actual conditions when the two I-V curves in figure 4 were measured.

A 10 point moving average was applied so that a clear correlation can be seen. By fitting a 3rd order polynomial fit, a functional relationship between FF and WUF is observed. The FF of the device is an indication of the series and junction quality of the device cells; therefore by plotting the FF with WUF a functional relationship can be established. Figure 8 shows the slight increase in FF as the WUF varies.

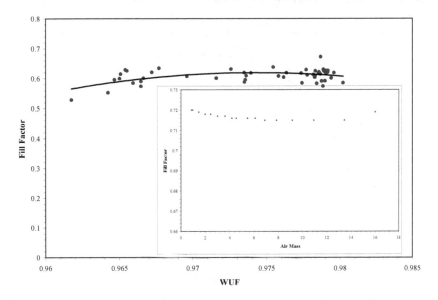

Fig. 8. Effect on CIS average Fill factor due to outdoor irradiance and spectral changes. Inset is the variation of FF vs. Air Mass for the same device.

Observed from figure 6, a 6.5% increase in FF is observed within the WUF range 0.960 - 0.983 (considering the % difference between the averages of the upper and low values of the FF). It should however be noted that this percentage increase value is just an indication of the change in FF. The increase in FF as observed is attributed to the quality of the spectrum dominating which result in 'supplying' sufficient energy for the electron-hole creation, with less energy losses, which in most cases is dissipated as heat. From the inset figure, a decrease in FF as AM values increase from AM 1.5 is evident. Closely analyzing the two graphs, the spectrum dominating under the WUF range of the CIS module is a blue rich spectrum which explains a slight increase in FF. From the inset figure, the FF is higher at AM 1.5 and decrease as the spectrum becomes longer wavelength dominated. Clearly the change in outdoor spectrum has an effect on the FF of the CIS module. Often reported is the relationship between efficiency and global irradiance as measured by the pyranometer. For CIS module, the variation of aperture efficiency with WUF is visible described by a logarithmic fit into the scattered data. Both WUF and irradiance affect device performance with the same magnitude. Gottschalg et al., (Gottschalg et al., 2004) established a relationship for device aperture efficiency and Useful Fraction (UF). The efficiency is described by $\eta \approx \dfrac{\alpha}{A}UF$ which when interpolated to our concept of Weighted Useful Faction

(WUF) the device efficiency would be described by $\eta \approx \dfrac{\alpha}{A}WUF$: where $\alpha = Power(P)/Spectral\,Re\,sponsiveRange(UI)$, is roughly a constant. This relationship exhibit a linear trend of efficiency with WUF in our case. The other key performance indicator in PV analysis is the device aperture efficiency. The efficiency of CIS module was also analyzed using the same procedure for FF analysis. Figure 9 indicate the efficiency versus WUF of the CIS device.

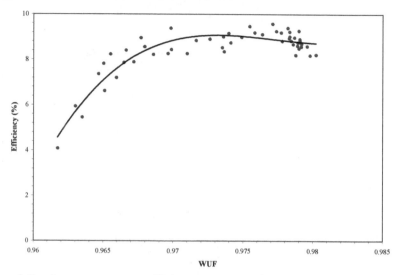

Fig. 9. Correlation between aperture efficiency versus outdoor WUF of the CIS module.

The efficiency increases logarithmically with an increase in Weighted Useful fraction (WUF > 0.960), which do not agree with the theoretical relationship illustrated in the previous section ($\eta \approx \frac{\alpha}{A} WUF$). One can attribute this discrepancy of the measured data and theory as follows: The α in the equation above is assumed to be a constant, but in actual fact it is strongly dependant on the irradiance available within the denominator function (UI). The irradiance within the Responsive Spectral Range (UI) is assumed to be a constant, a single value to be precise. In reality the irradiance does fluctuates within this range, rendering the α not to be a constant parameter. However the device efficiency exhibits a logarithmic increase as a function of WUF, due to the irradiance variations, resulting in α not to be a constant. The effect of season on device efficiency was also investigated; the results are shown in figure 10.

It is observed from figure 10 that the device efficiency is stable for both summer and winter. The PV module's performance parameters e.g. I_{sc}, V_{oc}, FF and η are characterized by what is referred to as temperature coefficients. Temperature coefficient is described as the rate of change (derivative) of the parameter with respect to the temperature of the PV device performance parameters (King et al., 1997). For PV system sizing and design, knowing the device temperature coefficient plays a very critical role. Quantifying the spectral effects on its own has proved to be a challenge; as a result no temperature coefficient with respect to outdoor spectrum has been documented. In figures 11 and 12, the relationship between outdoor spectral effects (WUF) and the average back - of module temperature is presented. Using a linear fit to the data, a spectral temperature coefficient is obtained. Figure 11 illustrates the relationship between WUF and temperature for a winter period.

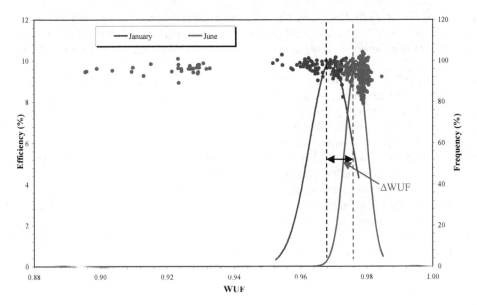

Fig. 10. Average outdoor aperture efficiency as a function of WUF of CIS module for both winter and summer period.

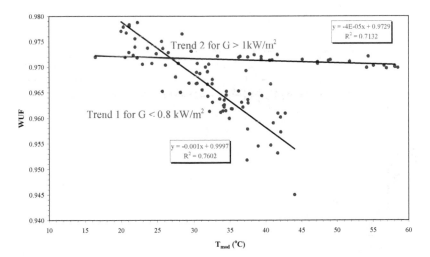

Fig. 11. Relationship between the outdoor WUF and back of module temperature of the CIS module during winter period.

Observing the results in figure 11, two temperature coefficients for WUF are obtained during the winter period. This trend in behavior could have been attributed to the different outdoor weather patterns observed for winter period. Some days even during winter, the outdoor climatic conditions would resemble a typical clear sky summer season, indicated by very high temperature (indicated by trend 2), while the rest of the days would be for typical winter season, normally characterized by mostly low temperature. In both cases, a negative WUF temperature coefficient is observed, with trend 1 being -0.001/°C and for trend 2 being -0.4×10⁻⁵/°C.

The same procedure was also used to find the effect of temperature on WUF for summer months of CIS module. Figure 12 shows the WUF versus temperature relationship.

Interesting to note from figure 12 is that the spectral effect temperature coefficient for summer period is the same as the one obtained during winter, clear sky (trend 2) although for summer the highest temperature reached was above 60°C while for trend 2 (figure 11), the highest was less than 60°C. From the two figures, it has been shown that temperature coefficient due to spectral effect (WUF$_\beta$) can be obtained once the outdoor spectrum data for a device is correctly calculated using the Weighted Useful Fraction (WUF) concept. Like other performance parameters, whose temperature coefficients are equally important in PV characterization and system design, the WUF should be also be considered as this might help to minimize some of the system sizing errors, which in most instances lead to under performance, unreliable and financial repercussions.

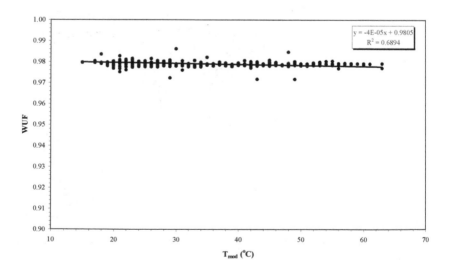

Fig. 12. Relationship between the outdoor WUF and back of module temperature of the CIS module during summer period.

5. Conclusion

The outdoor spectral effects using the Weighted Useful Fraction (WUF) of CIS module was analyzed. Observed was a 17% decrease in the device short - circuit (I_{sc}) current attributed due to a change in season. The change in season (summer/winter) result in the outdoor spectrum to vary by $\Delta WUF = 1.5\%$, result in the decrease in the device I_{sc}. From the analysis done, it was concluded that a small percentage change in ΔWUF resulted in large % difference of the module's I_{sc} as the outdoor spectrum changed during the course of the day, which confirmed that the 17% decrease in I_{sc} was due to a ΔWUF of 1.5 %. A strong correlation between FF and the WUF exists for CIS module. It is observed that the FF increases by 6.5% as WUF increases. The temperature coefficient of a device is one of the important concepts for characterizing device performance parameter. A close correlation between WUF and temperature was established. Temperature coefficients for spectral induced effect (WUF) were found to be $-0.001/^{\circ}C$ for winter period and $-4\times10^{-5}/^{\circ}C$ for summer seasons. This difference in WUF_β for summer and winter indicated that the temperature coefficients obtained in controlled environment (indoor procedure) can not be truly dependable for modeling purposes or system sizing since the outdoor conditions has an effect also. It should also be noted that the temperature coefficient for spectral effect is indeed an important parameter to consider.

6. References

Christian NJ, Gottschalg TR, Infield DG, Lane K (2002). Influence of spectral effects on the performance of multijunction amorphous silicon cells. *Photovoltaic Conference and Exhibition*, Rome

Gottschalg TR, Infield DG, Lane K, Kearney MJ (2003) Experimenatal study of variations of solar spectrum of relevance to thin film solar cells. *Solar Energy materials and solar cells*, vol 79, pg 527 – 537.

Minemoto T, Toda M, Nagae S, Gotoh M, Nakajima A, Yamamoto K, Takakura H, Hamakawa Y (2007). Effect of spectral irradiance distribution on the outdoor performance of amorphous Si//thin-film crystalline Si stacked photovoltaic modules., *Solar Energy Materials and Solar Cells*, Vol.91, pp. 120-122

M Simon and E.L Meyer (2008). Spectral distribution on photovoltaic module performance in South Africa. evaluation for c-Si modules", *33rd IEEE Phovoltaic Specialist Conference, San Diego*, California, USA.

Meyer, E.L, (2002). On the Reliability, Degradation and Failure of Photovoltaic Modules. *University of Port Elizabeth, PhD-Thesis*, 74-77, 34-38.

King, D.L, Kratochvil JA (1997). Measuring solar spectral and angle-of-incident effects on photovoltaic modules and solar irradiance sensors. *26th IEEE Phovoltaic Specialist Conference,Anaheim,CA,USA*

Riordan C, Hulstrom R (1990). What is an Air Mass 1.5 spectrum. *20th IEEE Phovoltaic Specialist Conference, New York*, pg 1085 – 1088.

Poissant Y, Lorraine C, Lisa DB (2006) (http://www.cete-vareness.nrcan.gc.ca).

Gottschalg R, Betts TR and Infield DG (2004). On the importance of considering the Incident Spectrum when measuring the outdoor performance of amorphous silicon photovoltaic devices. *Measurement Science and Technology*, vol.15, pg 460-466.

M Simon and E.L Meyer (2010). The effects of spectral evaluation for c-Si modules", *Progress in Photovoltaic: Research and Application,* DOI:10.1002/pip.973.

Permissions

The contributors of this book come from diverse backgrounds, making this book a truly international effort. This book will bring forth new frontiers with its revolutionizing research information and detailed analysis of the nascent developments around the world.

We would like to thank Professor, Doctor of Sciences, Leonid A. Kosyachenko, for lending his expertise to make the book truly unique. He has played a crucial role in the development of this book. Without his invaluable contribution this book wouldn't have been possible. He has made vital efforts to compile up to date information on the varied aspects of this subject to make this book a valuable addition to the collection of many professionals and students.

This book was conceptualized with the vision of imparting up-to-date information and advanced data in this field. To ensure the same, a matchless editorial board was set up. Every individual on the board went through rigorous rounds of assessment to prove their worth. After which they invested a large part of their time researching and compiling the most relevant data for our readers. Conferences and sessions were held from time to time between the editorial board and the contributing authors to present the data in the most comprehensible form. The editorial team has worked tirelessly to provide valuable and valid information to help people across the globe.

Every chapter published in this book has been scrutinized by our experts. Their significance has been extensively debated. The topics covered herein carry significant findings which will fuel the growth of the discipline. They may even be implemented as practical applications or may be referred to as a beginning point for another development. Chapters in this book were first published by InTech; hereby published with permission under the Creative Commons Attribution License or equivalent.

The editorial board has been involved in producing this book since its inception. They have spent rigorous hours researching and exploring the diverse topics which have resulted in the successful publishing of this book. They have passed on their knowledge of decades through this book. To expedite this challenging task, the publisher supported the team at every step. A small team of assistant editors was also appointed to further simplify the editing procedure and attain best results for the readers.

Our editorial team has been hand-picked from every corner of the world. Their multi-ethnicity adds dynamic inputs to the discussions which result in innovative outcomes. These outcomes are then further discussed with the researchers and contributors who give their valuable feedback and opinion regarding the same. The feedback is then collaborated with the researches and they are edited in a comprehensive manner to aid the understanding of the subject.

Apart from the editorial board, the designing team has also invested a significant amount of their time in understanding the subject and creating the most relevant covers. They scrutinized every image to scout for the most suitable representation of the subject and create an appropriate cover for the book.

The publishing team has been involved in this book since its early stages. They were actively engaged in every process, be it collecting the data, connecting with the contributors or procuring relevant information. The team has been an ardent support to the editorial, designing and production team. Their endless efforts to recruit the best for this project, has resulted in the accomplishment of this book. They are a veteran in the field of academics and their pool of knowledge is as vast as their experience in printing. Their expertise and guidance has proved useful at every step. Their uncompromising quality standards have made this book an exceptional effort. Their encouragement from time to time has been an inspiration for everyone.

The publisher and the editorial board hope that this book will prove to be a valuable piece of knowledge for researchers, students, practitioners and scholars across the globe.

List of Contributors

M. Estela Calixto
Instituto de Física, Benemérita Universidad Autónoma de Puebla, Puebla, México

M. L. Albor-Aguilera, M. Tufiño-Velázquez and G. Contreras-Puente
Escuela Superior de Física y Matemáticas, Instituto Politécnico Nacional, México

A. Morales-Acevedo
CINVESTAV-IPN, Departamento de Ingeniería Eléctrica, México

Nicola Armani
IMEM-CNR, Parma, Italy

Samantha Mazzamuto
Thifilab, University of Parma, Parma, Italy

Lidice Vaillant-Roca
Lab. of Semicond. and Solar Cells, Inst. of Sci. and Tech. of Mat., Univ. of Havana, La Habana, Cuba

Antara Datta and Parsathi Chatterjee
Energy Research Unit, Indian Association for the Cultivation of Science, Jadavpur, Kolkata, India

Satoshi Shimizu
Research Center for Photovoltaics, National Institute of Advanced Industrial Science and Technology, Japan
Max-Planck-Institut für extraterrestrische Physik, Germany

Michio Kondo
Research Center for Photovoltaics, National Institute of Advanced Industrial Science and Technology, Japan

Akihisa Matsuda
Graduate School of Engineering Science, Osaka University, Japan

Fan Yang
Qualcomm MEMS Technologies, Inc., United States

Gregor Černivec, Andri Jagomägi and Koen Decock
University of Ljubljana, Faculty of Electrical Engineering, Slovenia
Department of Materials Science, Tallinn University of Technology, Estonia
Solar Cells Department, Ghent University – ELIS, Belgium

Jhantu Kumar Saha
Department of Functional Material Science & Engineering, Faculty of Engineering, Saitama University, Japan
Current address: Advanced Photovoltaics and Devices, (APD) Group, Edward S. Rogers Sr. Department of Electrical and Computer Engineering, University of Toronto, Canada

Hajime Shirai
Department of Functional Material Science & Engineering, Faculty of Engineering, Saitama University, Japan

Yasuhiro Abe, Takashi Minemoto and Hideyuki Takakura
Ritsumeikan University, Japan

H. Il'chuk, P. Shapoval and V. Kusnezh
Lviv Polytechnic National University, Ukraine

Bolko von Roedern
National Renewable Energy Laboratory, United States of America

Michael Simon and Edson L. Meyer
Fort Hare Institute of Technology, University of Fort Hare, South Africa

Printed in the USA
CPSIA information can be obtained
at www.ICGtesting.com
JSHW011422221024
72173JS00004B/642